The Sixth Extinction

Also by Richard Leakey and Roger Lewin
Origins
People of the Lake
Origins Reconsidered

By Richard Leakey
One Life
Human Ancestors
An Illustrated Origin of Species
The Making of Mankind

By Roger Lewin
Darwin's Forgotten World
Thread of Life
Human Evolution: An Illustrated Introduction
Bones of Contention
In the Age of Mankind
Complexity: Life at the Edge of Chaos

The Sixth Extinction

Patterns of Life and the Future of Humankind

Richard Leakey and Roger Lewin

ANCHOR BOOKS
DOUBLEDAY

New York   London   Toronto   Sydney   Auckland

An Anchor Book

Published by Doubleday

a division of Bantam Doubleday Dell Publishing Group, Inc.
1540 Broadway, New York, New York 10036

Anchor Books, Doubleday, and the portrayal of an anchor
are trademarks of Doubleday, a division of Bantam Doubleday Dell
Publishing Group, Inc.

The Sixth Extinction was originally published in hardcover by Doubleday in 1995.

The Library of Congress has cataloged the Doubleday edition as follows:

Leakey, Richard E.
The sixth extinction: patterns of life and the future of humankind / Richard Leakey
and Roger Lewin.—1st ed.
    p.   cm.
Includes index.
1. Man—Influence on nature.   2. Evolution (Biology).   3. Extinction (Biology)
I. Lewin, Roger.   II. Title.
GF75.L425   1995
304.2—dc20   95-18286
CIP

ISBN 0-385-46809-1

10 9 8 7 6 5 4 3 2 1

To our fellow species
and our collective future

Acknowledgments

Each of us owes a debt of gratitude to many people who have enriched our professional lives, which span a great range of experiences. Those people, too numerous to name in full, know who they are.

In particular, however, Leakey would like to express his thanks to the staff of the Kenya Wildlife Service, who supported his efforts and vision while he was director. He would also like to pay tribute to his wife, Meave, who helped him walk again after he lost his legs in a plane crash. He owes his life to her.

Lewin would like to voice his appreciation to just a few of those who were especially encouraging and inspiring to him in his quest for the patterns of life and their meaning, acknowledging that such a list is necessarily incomplete. They are: Stephen Jay Gould, David Jablonski, Thomas Lovejoy, Robert May, Stuart Pimm, David Raup, and Edward O. Wilson.

Contents

The Sixth Extinction

1

A Personal Perspective

N O ONE WAS MORE SURPRISED than I was when, one April afternoon in 1989, a colleague burst into my office at the museum, in Nairobi, and exclaimed excitedly, "Congratulations!" Puzzled, I replied, "On what?" My colleague told me he'd just heard on the radio that I'd been appointed director of the Wildlife Conservation and Management Department. The decision had come from the president, Daniel arap Moi, and had just been announced on the radio. "News to me," I said, and quickly left the museum—the day's work undone—and returned home. I spoke to the president on the telephone the following morning, and he asked me to meet him to discuss how I might launch an effort to transform wildlife management in our country, including preventing elephants from being driven to extinction in a lethal hail of poachers' bullets.

So began one of the most challenging periods in my life. It would take me away from my post as director of the National Museums of Kenya, which I'd held for two decades, and from my deep love of paleoanthropology, the search for human origins, which I had pursued for an equal length of time in the fossil-rich sediments of northern Kenya, on the eastern and western shores of Lake Turkana. It would also take me back to my roots.

When I was a young boy, the smell of the soft air, the sight of wild places and wild creatures, and the sounds of unseen animals in the night seeded deeply in me a love of Africa, a love of nature. At the time I was much more interested in living things than in old bones, which, to my puzzlement and frequent irritation, so fascinated my parents, Louis and Mary. Their discoveries, which established East Africa as a vital region for uncovering evidence of our earliest ancestors, are legendary in the annals of human origins research. In addition to his obsession with the past, however, my father was a passionate naturalist, and he wrote several books on animals of the region. He also founded the East Africa Wildlife Society, in 1958, which still plays an important role in ecological research and conservation in the country. When we were young, Louis would tell me and my brothers, Jonathan and Philip, endless stories as we walked through the African bush at Olduvai Gorge—he hoping to find clues to new fossil sites, we hoping for a glimpse of a lion kill or a leopard stalking. We were often rewarded with storybook sights. He also held us rapt with stories in camp at night, the still air full of nature's sounds. I became an ardent naturalist, too, and at first was especially entranced by beetles and butterflies. Later, I recognized the wonder of larger creatures, and I mused on the diversity of life and how each part of it interacted intimately with so many other parts. In 1969 I founded the Wildlife Clubs of Kenya, whose purpose is to educate children about life in their land.

I love Kenya, land of my birth. It is a country of tremendous physical contrasts, from the hot and humid coastal habitats at sea level to the snow-clad peaks of Mount Kenya (Africa's second highest mountain, after Kilimanjaro); from arid desert to moist mountain slopes. The diversity of life thriving in these different habitats matches the physical diversity of the land, and is among the richest in the world. As a boy, I wasn't familiar at the time with the phrase "the balance of nature," but it captures in a simplistic way what I resonated with in the wild.

I followed my urge to be in wild places as much as I could, and still do, because nature was and still is cleansing and recharging to what, for want of a better word, I call my soul. Raw nature has its dangers, too, of course. Both my brothers and I suffered from frequent bouts of malaria and occasional outbreaks of bilharzia, a parasitic infection contracted from bathing in water containing snails harboring *Schistosomiasis mansoni*. There were snake bites, too, which were usually more dramatic in appearance than life-threatening, though not always so. And on one ignominious occasion I had to lock myself in a cage I had set to catch a leopard, which seemed determined to pay more attention to me than I thought healthy. Despite the humiliation, it seemed prudent to be my own temporary prey than the cat's permanent victim.

As a young teenager, I fantasized about being a game warden one day, but contented myself with capturing animals for the filmmakers Armond and Michaela Denis, who lived near my parents' house in a Nairobi suburb. For many British television viewers, the Denises' films were their first introduction to wild Africa. In fact, the Denises often staged close-ups using the animals I had caught for them. For the time, it was a legitimate technique, given the limitations of getting close to dangerous animals in the wild. In capturing animals, I learned a lot about their behavior. I was also paid for the animals I delivered, so my education as a naturalist was combined with building what, for me at the age of thirteen, was a healthy bank account. I liked the beginnings of independence it gave me.

Nature and commerce conspired again in my education, this time through the agency of a severe drought in 1960 and 1961. Tens of thousands of animals succumbed, and the plains were littered with their corpses, too numerous for the scavengers to dispose of. Many corpses lay in pristine condition, unmarked by tooth or beak. Having declared I would become financially independent of my parents (I was seventeen at the time), I saw the bounty of harsh nature as an opportunity for me. I borrowed money to buy an old Land-Rover, and set off to collect dead animals, small and large. I boiled the soft tissues off them in an old oil drum, cleaned and dismantled the skeletons, and shipped them to museums and universities the world over, for a satisfying profit. In the process I became intimately acquainted with comparative anatomy, because I had to label and number each bone so that the skeletons could be reassembled at their destination. I can think of no more effective means of learning anatomy. As a paleoanthropologist you often have to identify an animal from mere fragments of its fossilized bones. So, although I didn't know it at the time, my brief, early career as a bone peddler provided me with a solid foundation for my later, lengthy career as a paleoanthropologist.

Before I became deeply involved in human origins research, however, I formed my own safari company, which was a wonderful opportunity for being in remote, wild places, and being paid for it. It thrilled me to introduce visitors from Europe and America to the great diversity of life in my own country, ranging from the tiniest detail of an orchid flower to the great migrations of wildebeest from the Serengeti in Tanzania to the Masai Mara in Kenya. I was in my element and was tremendously happy. But I was also restless; I knew that I wanted to do something else, but did not know what that something was. Eventually, despite strident vows that I would never step into the professional footsteps of my parents (or, more accurately, fall into their shadow), I became a paleoanthropologist, and launched my first major expedition to the east side of Lake Turkana in 1968. I have never regretted the decision, because I have been lucky to work with some fine scientists and have had the opportunity to discover prized relics of our evolutionary history. Many people experience a deep, almost primordial urge to understand our beginnings as a species, and the search for such relics in ancient sediments brings one into direct contact with our species' history. Those of us who are in this line of work are truly privileged.

For two decades I combined my job as director of the National Museums of Kenya—a series of ten museums throughout the country—with spending as much time as I could manage in the field, looking for and excavating fossils. The eastern and western shores of Lake Turkana have proved to be wondrously rich sources of early human fossils, revealing our evolution from some four million years ago to relatively recent times. The story of our evolution is now much more complete than it was twenty years ago, and I'm proud to have contributed to some of that greater understanding through spectacular finds on both sides of the lake. When you are walking over ancient sediments, seeking and finding fossils, you are doing more than simply picking up old bones, important though some of them may be. You are doing more, too, than reconstructing the evolutionary history of a particular species, *Homo sapiens*. You are looking through a paleontological window onto past worlds, witnessing their fate through time.

If there is one single impression you gain from what is to be seen through this window it is encapsulated in the simple word *change*. Life's flow is in a constant, dynamic change. Sometimes it is driven by a shift in climate, so that terrain that once was arid may become moist, bringing with it a shift in the cast of characters that are able to live there. Sometimes it is driven by a burst of evolutionary turmoil, so that creatures that once existed are no more, and new ones take their place. Bouts of extinction and speciation are a periodic force in changing life's flow, generating constantly shifting patterns. Life seen through a paleontological window is like a kaleidoscopic image, where change is not only natural but inevitable. You see death as part of life, extinction as part of life's flow.

When I acceded to President Moi's behest to become director of the Wildlife Conservation Department, I was faced with some very immediate practical issues, not least of which, as I mentioned, was the urgent problem of putting an end to the avaricious business of elephant poaching. There was also the challenge of reconciling the conflicting needs of a growing human population, which requires ever more land, and the protection of wildlife, whose natural habitat was being encroached upon. But I was also able to understand the nature of diversity of life, and the place of *Homo sapiens* within it, by viewing it through the perspective of constant change that is an inevitable aspect of Earth history. I'm not suggesting that such a perspective helps much when faced with what to do, say, about the damage to crops that sometimes occurs when elephants stray into farming communities. But it is helpful in reaching a philosophy of how one should react, say, to the interaction of elephant populations and habitats they sometimes seem to be destroying by their presence. I developed the conviction that, often, it is best to allow nature to take its course, that what we are witnessing in such a situation is the process of change that is part of nature and that it is futile—harmful, even—to try to prevent it. I'll return to this issue later in the book. Most important of all, however, the perspective of change throughout the history of life offers us a means of judging our rights and responsibilities as a species, and the rights of other species with which we share the Earth.

When I started to think of the kind of book I wanted to write, I realized that my experience as paleontologist and conservationist offered a unique view on our current predicament. This is not the first book that argues that *Homo sapiens,* dominant species that it has become, may be in the process of causing a biological catastrophe of immense proportions, through eroding the diversity of life at an alarming rate. (For instance, the felling of tropical forests and the encroachment on wild places through economic development may soon be pushing as many as 100,000 species into extinction each year.) But it is the first to address the phenomenon from the perspective of *Homo sapiens* as but one species in a flow of life that has a long history and a long future. In order to know ourselves as a species and to understand our place in the universe of things, we have to distance ourselves from our own experience, both in space and time. It is not easily done, but it is essential if we are truly to see a larger reality. It is a perspective that should make us humble, particularly given the tremendous power we now wield to alter our planet in profound ways.

Several themes run through this book, of which the notion of change is central. From it we recognize that humans are but a brief moment in a continuous flow of life, not its end point. But there is more to be learned from change than the place of humans in the world. Most important, it is in the *patterns* within the change that we find the nature of life's flow; the patterns are the surface signals of the fundamental processes that nurture that flow. What do I mean by patterns? I mean the images that emerge when we scrutinize the fossil record in its entirety. I mean the images that emerge when we scrutinize ecological communities as a whole. Each of these images is composed of individual entities, of course: the fossilized remains of individual species in the geological record, for instance, and the individual living species that make up ecosystems. But it is in the relationship among species, in present and former communities, that we find the true nature of the world we live in.

Striving to see these images is like seeing the three-dimensional shapes that emerge from an apparently meaningless scatter of dots in a Magic Eye picture. You can look at these pictures for a long time, and see nothing but the dots. But suddenly, when your mind is receptive and you stop focusing on the surface pattern, you see beyond it and you recognize a deeper visual reality. The sciences of evolutionary biology and ecology are at the brink of recognizing that deeper reality in life's flow. The images are still incomplete, but they are sharp enough for us to be able for the first time to see a new reality in the world, and it represents nothing short of an intellectual revolution. Ours is a very different world from what we thought it was just a few years ago.

It is now possible to look at the pattern of mass extinctions in the fossil record and see these events as a major creative force in shaping life's flow, not simply as occasional interruptions of it. That's new. We can look at the deep processes of evolution and see that all of life, including *Homo sapiens,* is something of a grand lottery. That's new. And we can now look at the pattern of ecological communities and see how they assemble themselves and how unexpected dynamics emerge from them. That's new, too. These novel insights from evolutionary biology and ecology are combining to cut through the deceptively simple and unimaginably complex phenomena that constitute the living world around us.

There's a popular misconception that among the sciences the greatest intellectual challenges are to be found in the physical sciences: physics is "hard" science; biology is "soft"; or so it is often said. In fact, the living world and its history encapsulated in the fossil record are incredibly complex, and a full understanding of them still eludes us. I used the phrase "the balance of nature" earlier, a much-used phrase that seems to reflect the simplicity and harmony of life. It's wrong, as we will see in later chapters. Nature is not simple, and the supposed harmony is wildly misleading.

Homo sapiens shares this world with millions of other creatures, together constituting an awe-inspiring diversity of complex life. My aim in the following pages is to use the new insights from evolutionary biology and ecology to create a greater appreciation of that diversity and its fate. We need to understand the *source* of that diversity and its extent. Why, for instance, should there be, say, fifty million living species in the world today, not one million or five hundred million? We need to understand the *place of humanity* in that diversity. Are we an inevitable product of the flow

of life, and its pinnacle? We need to understand *human impact* on
that diversity. Is our species capable of destroying millions of
species? If so, how does it happen? And we need to understand
the *future* of that diversity. From the patterns of the past, can we
predict what will happen in the future?

When I give talks about human origins research, the question
I'm most frequently asked is "What will happen next?" I understand the concern that is the source of this question. It is a perva-
sive uncertainty about the future of humankind, and the ques-
tioners are usually seeking some kind of reassurance. The
answer, which I will develop in the following pages, is often not
welcomed, because it gives no such reassurance.

As with several of my previous books, I collaborated with Roger
Lewin on this one. We combined our different viewpoints in creating this journey through the history of life, in search of patterns that reveal its nature and future: I draw on my experience
as a paleoanthropologist and conservationist, and Roger on his
expertise in evolutionary biology and ecology. It has been a
shared journey, but, as has been our practice in other collabora-
tions, the text is written in my voice. The decision to do it this
way again reflects in part my own role in promoting issues of
conservation on an international stage, and is in part a conve-
nient literary device. Also, the fact that we are speaking in one
voice is symbolic of our shared vision of the world of nature and
our shared concern for our fellow species.

Time and Change

LOCKED AS WE ARE in our world of the present, it is a challenge for us to appreciate the flow of evolutionary processes that shaped the world we inhabit.

Homo sapiens is but one species among many, product of an intricate and unpredictable interplay between the creative processes of evolution and the sometimes capricious hand of extinction.

In this section we will glimpse some of the fundamental features of that creativity—and its persistent mysteries— and see the overriding importance of occasional crises in the history of life.

2

Life's Salient Mystery

O N THE WESTERN SHORE of Lake Turkana, not far from where in 1984 we excavated the skeleton of a youth who had died 1.6 million years before, is an eerie sight. Looking like petrified hippos wallowing in a lake of dry sediment, hundreds of fossil stromatolites in orderly array betray the ancient presence of a shallow lagoon. I saw these strange "creatures" for the first time while taking a break during the early days of the excavation of the skeleton, which we soon came to call the Turkana boy. Frank Brown, Alan Walker, and I had left our colleagues still working under the heat of the tropical sun, and had driven westward, away from the lake, for almost a mile. Alan, an anthropologist at The Johns Hopkins University, has been a close friend and coworker for years, and we've unearthed many ancient human relics together. Frank, a geologist from the University of Utah, has spent a decade rolling back the geological history of the lake basin, revealing the many forces that have shaped the region. He had recently found the stromatolites, and he wanted us to see them; hence our small expedition from the excavation camp. Eventually, we reached a water course adjacent to the Nariokotome River to the north, where the boy's skeleton lay. Both rivers are dry as dust at that time of year, August.

By now, we were some three miles from the shore of the lake, and I could see, stretching into the distance, the rounded, astonishingly regular form of the ranks of stromatolites. They lie on a virtually barren plain below a ridge strewn with lava boulders, high and dry. But, as Frank explained, a million years ago they were covered by shallow, still waters. The lake evidently had been much bigger then. When alive, stromatolites are colonies of single-celled algae and other microorganisms, engaged in a complex microecology. The colonies are formed as thin layers of microorganisms, centered initially, perhaps, on a grain of sand. Slowly, they grow as fine sediment covers the outer, living layer, forcing the now-buried organisms to struggle once again to the surface, once again to form a thin, flat, complex microbial community. In stromatolites, life is always clinging to the edge. The results are structures the shape of flattened spheres, some of which reach the size of a small table, three feet in diameter.

Some of the Lake Turkana stromatolite fossils had sundered in two, revealing, layer upon layer, the signature of successive colonies, life on a microscopic scale. Their presence in this terrain gave us clues to the conditions in the Lake Turkana basin a million years earlier: we could see that lake waters once lapped where now was arid desert. They also gave us a view of life much deeper into the past, as deep, in fact, as it is possible to peer and still see life. Although living stromatolites are rare in today's world—the salty, hostile water of Shark Bay, in Western Australia, is one of the few places where they still thrive—they are common in the fossil record and are among the first manifestations of life on earth.

One of the most astonishing facts about Earth history is that life arose so very early. The planet condensed from the debris of the nascent solar system 4.6 billion years ago, a searing agglomeration of molten, radioactive rock, quite inimical to any form of life. Slowly, the heat of its fiery birth dissipated, so that a little less than four billion years ago life became theoretically possible; that is, fragile organic molecules were no longer wrenched apart as soon as they formed, and watery environments could persist rather than being blasted into vapor. Both conditions were necessary for life to emerge. That theoretical possibility was quickly followed by reality, as primordial life dawned, in the form of simple, single-celled organisms, ghosts of which have been found in the oldest known continental rock, some 3.75 billion years of age. Harnessing the energy of the sun, and exploiting their chemical environment, these simplest of organisms, cells without nuclei—or prokaryotes, as they are known—proliferated and diversified in type. Collectively, they formed the layered mats of microorganisms that, through time, created the characteristic colonial form of the stromatolite.

With so early a start, life might have been expected to immediately set on a course of gradual and steady progression toward ever more complex forms: first, more complex cells, or eukaryotes, in which genetic material is packaged in nuclei, became established eventually; specialized organelles—mitochondria and chloroplasts—that perform special functions, arose; then, onward to multicellular organisms, simple at first, but marching from invertebrates to vertebrates, from amphibians and reptiles to mammals—and, ultimately, to us, *Homo sapiens*. From the vantage point of the present we can see that the stages of life I just mentioned indeed occurred, but they did so in what can only be described as an erratic and unpredictable way. If there is one thing we learn about life by studying its history on Earth, it is that little about it is gradual and steady. (This applies on all scales, from global to local, like fractal patterns.)

After its urgent first foothold, the most complex form of life for a mind-cheating two billion years remained the prokaryotic cell, and its most complex organization the microecology of stromatolite colonies. Life, it seemed, was in no hurry to go anywhere. When eukaryotic cells finally did appear, some 1.8 billion years ago, the scene might seem to have been set for a race to the next stage, that of multicellular organisms. But no; another billion years and more were to pass before such organisms evolved. Even when they did materialize, such organisms were cryptic at best, and were not obviously the harbingers of the complexity of organization and interaction we mean when we think of multicellular life. The advent of complex multicellular organisms—by which I mean rather modest, if bizarre, marine invertebrates— had to wait until some 530 million years ago, or after 85 percent of current Earth history had been chalked up. But when it began, it did so spectacularly, as an event that paleontologists call the Cambrian explosion.

In the space of a few million years, all the major body plans, or phyla, that represent life on the planet today were invented in a frenzy of evolutionary innovation. Among them was a tiny organism, which looked like the modern marine wormlike creature *amphioxus,* blessed with the scientific name *Pikaia,* which was the probable founder of the phylum chordata, which includes all

Biological diversity evolved rapidly during the Cambrian explosion, 530 million years ago. Shown here is a reconstruction of life found in the Burgess Shale. They include some familiar forms (such as sponges and brachiopods) that have survived to the present. But many (such as the giant Anomalocaris, *lower right) left no descendants and represent phyla that no longer exist. (From "The evolution of life on Earth" by Stephen Jay Gould. Copyright © 1994 by* Scientific American, *Inc. All rights reserved.)*

Amphioxus, a marine creature alive today, is very similar to Pikaia, at least in the level of anatomical organization.

later vertebrates, including *Homo sapiens.* From the vantage point of the present, it was a modest start in the extreme.

"Unprecedented and unsurpassed," is how James Valentine, a paleontologist at the University of California in Berkeley, has described the Cambrian explosion. It is an apt, even understated, phrase. And, as we will see in the following chapter, there is even more to the Cambrian explosion than its explosiveness.

The arrival of complex multicellular life, late in the day though it was, again might have been expected to presage steady progression through the prehistoric worlds we know from the fossil record, leading eventually and predictably to the world of nature as we know it today. But, again, such expectations were not met. Life since the Cambrian explosion has been characterized by boom and bust, as species diversified wonderfully, only to be extirpated in huge numbers by occasional mass extinctions, five in all. The epigram that is said to describe war is fitting for life on Earth, too, in its tempo if not in its manifestation: long periods of boredom interrupted by brief moments of terror.

One of the legacies of the Darwinian revolution, which a century ago replaced the traditional explanation of life as divine design with a naturalistic perspective, imposed a very particular view of the world on Western intellectual thought. According to that perspective, species thrive because they are superior in some way to their competitors; they win in the "struggle for existence," to use Darwin's phrase. Similarly, species go extinct through succumbing to competition; they are failures in life's struggle. This is a rather simplistic reading of Darwin's greatest insight, the theory of evolution by natural selection, but it was comfortingly consistent with the Western ethos of success through effort. The expression of this ethos can be seen, sometimes explicitly, sometimes implicitly, in biological texts and even more so in anthropology texts, particularly those of a few decades ago.

The biological diversity of life today—the numbers of species in their different habitats—is close to its peak in Earth history. We are one species among uncounted millions, the richest tapestry of life the planet has ever hosted. With the perspective of "evolutionary success through superiority" I just described, we could feel justified in patting ourselves on the back and counting *Homo sapiens* as being among the collective product of the steady success of species ever more suitably adapted to their environments. But one of the most important developments of evolutionary biology in recent years is that luck, not superiority, plays a cogent role in determining which organisms survive, especially through times of mass extinction. We therefore have to accept that humans are in the company of the lucky survivors of cataclysmic convulsions in Earth history, not the modern manifestations of ancient superiority.

Biologists are interested in the patterns of life on Earth and in what drives those patterns. Namely, how do new species arise, and under what circumstances do they disappear? In other words, what drives those patterns? The interplay of the dynamics of the origin and extinction of species determines the diversity of life on Earth at any one point in history; prevailing biological diversity is the product of the past and it sets the stage for the future. This section, Time and Change, of our book will explore the biological foundation of the rich diversity of today's world, namely, the Cambrian explosion; it will describe the gauntlet of mass extinction through which life has passed on its way toward the present; and will focus on the place of ourselves, *Homo sapiens,* amid today's rich diversity of life.

Charles Darwin was puzzled and disturbed by the Cambrian explosion, because he believed that the sudden appearance of many groups of species posed a serious challenge to his embryonic theory of evolution by natural selection. The essence of natural selection, as Darwin conceived it, was gradual change, the accumulation of tiny modifications in behavior or anatomy as a response to prevailing environmental circumstances. Over very long periods of time, a substantial evolutionary shift could result, sometimes yielding new species, but, Darwin believed, the journey there must be slow. How, then, was the apparently explosive appearance of life in the earliest part of the record to be explained?

GEOLOGIC ERAS			
Era	Period	Epoch	Approximate number of years ago (millions of years)
Cenozoic	Quaternary	Holocene (Recent) Pleistocene	
	Tertiary	Pliocene Miocene Oligocene Eocene Paleocene	
Mesozoic	Cretaceous Jurassic Triassic		65
Paleozoic	Permian Carboniferous (Pennsylvanian and Mississippian) Devonian Silurian Ordovician Cambrian		225
Precambrian			570

The geological time scale.

Darwin sought solace in the imperfect fossil record, a subject to which he devoted an entire chapter in *The Origin of Species*. He argued that the Cambrian explosion looks dramatic, only because the ancestors of these organisms had yet to be discovered. In Darwin's time, no fossil evidence had been uncovered of life earlier than the Cambrian, a fact he found "inexplicable."[1] Perhaps, he suggested, Precambrian organisms had failed to be incorporated into the fossil record or, through later geological events, had been erased from it. He even speculated that the relevant fossils had accumulated on continents that were now deeply submerged beneath the oceans and therefore beyond the paleontologist's reach. (Although modern geological theory precludes the disappearance and appearance of continents in such a way, a century ago such processes were held possible.)

Every paleontologist knows that, as Darwin mourned, the fossil record is often frustratingly incomplete, for a host of geological reasons. Those of us who are interested in human prehistory would dearly like to find good geological exposures in the four- to eight-million-year range in East Africa, because we know that interesting developments took place in that interval. Alas, they remain tantalizingly elusive. Darwin's appeal to the imperfection of the record, therefore, was not a resort to desperation, but an entreaty to an unfortunate geological reality. One day, he said, evidence would be unearthed that would expose the long prelude to the apparently explosive events of the Cambrian. A gradual, progressive trajectory of change would then be apparent, smoothing out the (to Darwin) uncomfortably abrupt emergence of complex forms of life.

Almost a century passed before evidence of life in the Precambrian came to light.

In 1947 the Australian geologist R. C. Sprigg made a historic discovery when he found creatures resembling jellyfish in ancient sediments of the Ediacara Hills of the Flinders Range, in South Australia. The significance of the discovery lay in the age of the deposits, some 670 million years old, which predates the Cambrian explosion by more than a hundred million years. Eventually, further expeditions added other organisms to the list, including creatures that were said to be similar to other jellyfish, segmented worms, arthropods, and corals, animals we are familiar with in today's world. Some looked like no living creature ever seen, extant or extinct.

All of these organisms, known collectively as the Ediacaran fauna, were soft-bodied; that is, they lacked calcified shells. One of the explanations that had been advanced to explain the apparent lack of precursors to the denizens of the Cambrian world was that they were soft-bodied and therefore highly unlikely to become fossilized. Only under the most extraordinary geological circumstances would it be possible to snatch brief glimpses of this fragile Precambrian life. The fine sandstone sediments of the Ediacara Hills apparently matched those conditions, preserving the shapes of these creatures rather than crushing them out of existence.

Since that first discovery almost half a century ago, similar assemblages of primitive soft-bodied creatures from the same time in the Precambrian have been found in many parts of the world, removing the possibility that the Ediacara Hills' trove was a geological freak. The Ediacaran fauna were varied and world-wide, elements of an apparently important stage in the progression from simple single-celled organisms to more complex single-celled organisms, and eventually to the most complex life forms, that of multicellular creatures. That history, the progression from simple to complex organisms, was, however, not an inexorable, predictable process. As we've seen, the arrival of a multicellular world was late in Earth history, the probable reason for which has recently been discerned. "The emergence of [multicellular] animals was closely linked to unprecedented changes in the earth's physical environment, including a significant increase in atmospheric oxygen," explained Andrew Knoll, a geologist at Harvard University.[2] Until oxygen levels in the atmosphere rose from a lowly 1 percent to something close to the 21 percent of today's world, organisms larger than a single cell simply could not survive. The idea of an oxygen barrier to the emergence of complex forms of life has a long pedigree, going back three decades, but geochemical evidence of a sudden increase in atmospheric oxygen was discovered only recently.

Modifications in the physical environment can be a powerful engine of evolutionary change, and has often been *the* critical factor. This was the case, I believe, with the origin of the human family, a recent event of some five million years ago; it may have been instrumental in the origin of our own genus, *Homo;* and it seems to have been the case with the first appearance of multicellular life, some 630 million years ago. Such an explanation would have appealed to Darwin, because he was always aware that organisms' "struggle for existence" involved an interaction both with other forms of life and with the physical environment.

The discovery of the Ediacaran fauna elicited a collective sigh of relief from the geological community, which, for the most part, held to Darwin's view of gradual development through long periods of time. Here, at last, were the ancestral fauna to life in the Cambrian world. The Cambrian explosion wasn't an abrupt appearance after all. It was, as Darwin predicted, merely a mischievously misleading artifact of a notoriously incomplete fossil record. The "vast, yet quite unknown, periods of time" prior to the Cambrian world had indeed "swarmed with living creatures," just as Darwin had written.[3] Although the long prelude of single-celled life could not be doubted, the origin of more complex creatures could now be seen to be gradual, not abrupt.

Several decades were to pass before doubts fermented over the true nature of the Ediacaran fauna. Although some of the creatures were mysterious and could not readily be linked with any later form of life, most had been identified as being ancestral to one type of Cambrian animal or other. Since the mid-1980s, however, most of those links have been questioned, with the possible exception of sponges. Adolf Seilacher, a paleontologist at the University of Tubingen, Germany, was among the first to challenge the conventional interpretation, and certainly the most influential. Seilacher acknowledged superficial resemblances between Ediacaran animals and later species, but argued that the fundamental architecture was different. Modern organisms handle the transport of nutrients and respiratory gases via various different systems of internal tubes. No such systems existed in the Ediacaran fauna, which possessed "a unique quilted-pneu construction,"[4] observed Seilacher. Because the internal structure of the Ediacaran animals was so radically different from later organisms, Seilacher argued that they could not be ancestral to the Cambrian bestiary. The Ediacaran fossils, he said, "represent a failed experiment of Precambrian evolution rather than the ancestors of modern animals and plants."[5]

Seilacher's view soon prevailed. For instance, in October 1989 Simon Conway Morris, a geologist at Cambridge University, England, wrote in the journal *Science* about "a striking lack of continuity [between the Ediacaran fauna and] the succeeding Cambrian fauna." Although he noted possible geological artifacts that might cause this disparity, he concluded that the discontinuity "may reflect major extinctions in the Ediacaran assemblages."[6] And in the journal *Nature,* four years later, he elaborated on this conclusion and presented a diagram of life in

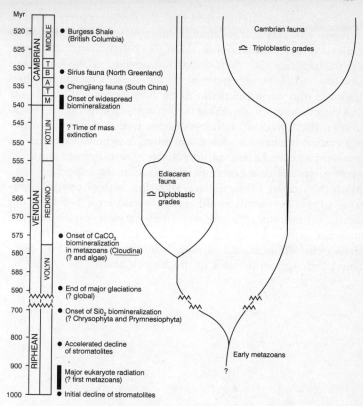

Myr				
520	CAMBRIAN	MIDDLE	● Burgess Shale (British Columbia)	Cambrian fauna
525		T		⇆ Triploblastic grades
530		B	● Sirius fauna (North Greenland)	
535		A / T	● Chengjiang fauna (South China)	
540		M	Onset of widespread biomineralization	
545	VENDIAN	KOTLIN	? Time of mass extinction	
550				
555				
560		REDKINO		Ediacaran fauna
565				⇆ Diploblastic grades
570				
575			● Onset of CaCO₃ biomineralization in metazoans (Cloudina) (? and algae)	
580		VOLYN		
585				
590			● End of major glaciations (? global)	
700			● Onset of SiO₂ biomineralization (? Chrysophyta and Prymnesiophyta)	
800	RIPHEAN		● Accelerated decline of stromatolites	Early metazoans
900			Major eukaryote radiation (? first metazoans)	?
1000			● Initial decline of stromatolites	

The Ediacaran fauna, once thought to be ancestral to all later forms of complex organisms, probably were a separate evolutionary experiment, leaving few if any descendants. (Reprinted with permission from Nature, *vol. 361, page 219, by Simon Conway Morris. Copyright © 1993 Macmillan Magazines Limited.)*

Precambrian and Cambrian times, showing the initial burgeoning of the Ediacaran fauna, followed by its almost total eclipse, the result of worldwide extinctions of creatures, the like of which has not been seen since.

For a hundred million years, the Ediacaran animals were the most complex forms of life on Earth, and their advent wreaked havoc among the existing communities of microorganisms. Prokaryotes, joined later by single-celled eukaryotes, had lived in colonies as life's highest form of organization for more than three billion years. Rates of the appearance of new species and of extinction had varied from time to time, with occasional periods of high species turnover, but never had there been any mass dyings. That changed when the exotic bestiary of the Ediacaran times dawned: some 75 percent of the species of single-celled organisms that constituted the living layers of stromatolites became extinct, victims of grazing by the new big boys on the block. But then the Ediacaran animals disappeared, too, virtually without a trace, victims of we know not what. It is now clear that for the most part the Ediacaran animals were ancestors of nothing, merely remnants of a grand, failed evolutionary experiment.

With the demise of the Ediacaran fauna as the hoped-for precursor to the Cambrian world, which would have provided the gradual evolutionary progression that Darwin predicted, the Cambrian explosion emerged once again as an enigma that required extraordinary explanation. It has with justification been termed "the salient mystery in the history of life."[7]

3

The Mainspring
of Evolution

T HE CAMBRIAN EXPLOSION half a billion years ago was a
burst of evolution unprecedented in the history of life and,
as we will see, has not been repeated since. A panoply of new life
forms arose in a geologically brief period of time. Biologists
would dearly like to know what propelled this salient mystery of
life. According to David Jablonski and David Bottjer, paleontologists at the University of Chicago, "Innovation is the mainspring
of macroevolution."[1] The term macroevolution refers to the major shifts in the history of life, such as the origin of flowering
plants and the origin of placental nourishment of embryos, a
component of the evolution of true mammals. Macroevolution
describes the larger patterns in the history of life, and there is no
larger nor more dramatic pattern than the Cambrian explosion,
when the evolutionary scene was essentially set for the rest of
Earth history. All the basic forms of architecture upon which
modern multicellular organisms are shaped came into being in
that brief burst of evolutionary innovation.

Until recently, that interval was judged to have lasted some twenty to thirty million years, which, measured on the usually leisurely scale of geological time, was brief indeed, especially given the magnitude of change involved. Late in 1993, however, a team of geologists and paleontologists based at Harvard University slashed the estimate for the duration of the Cambrian explosion by more than half, arguing that it lasted no more than ten million years, and perhaps as little as five million. My friend Stephen Jay Gould, also of Harvard University, commented at the time of the discovery: "Now this greatest of all evolutionary bursts is even more of an evolutionary burst than we thought."[2] Clearly, the gradual pace of canonical Darwinian evolution will not suffice as an explanation of an event of this magnitude and rapidity, and other mechanisms must be sought. Jablonski and Bottjer warn that biologists face a tough task in trying to explain how the mainspring of this extraordinary event might have functioned. They wrote, "The most dramatic kinds of evolutionary novelty, major innovations, are among the least-understood components of the evolutionary process."[3]

This chapter has two goals. First, it will explore the nature of the Cambrian explosion and will relate some of the hypotheses that have been proposed to account for it. Second, it will focus on the recently discerned, unusual character of the event, one that challenges traditional descriptions of Earth history.

When tourists come to Kenya to view wildlife they have at the top of their wish list elephants, lions, rhinos, leopard, and buffalo, some of the most conspicuous and romantic of African wildlife. People feel cheated if they return home without stories of being close to nature, seeing these beasts in the wild. I can understand this, for these majestic creatures strike awe in the viewer. They are spectacular and irreplaceable players in the rich ecosystem of my country, my continent. We humans appear to be drawn to the dramatic—and there's no question that a matriarchal group of elephants, lions at a kill, and a menacing herd of buffalo are dramatic. But they are only a part of Africa's ecosystem and of all ecosystems worldwide, and a rather small part, too. Mammalian species, the great majority of which are much smaller in size than these majestic creatures, number perhaps four thousand, only a tiny fraction of the many millions of species that constitute the Earth's current biota. We are entranced by a few of life's rarities, which live among and depend upon a vast kaleidoscope of "lesser" players in the game of life.

Vertebrates are chordates, one of about thirty modern animal phyla. Most of the remaining twenty-nine are what we would probably regard as modest forms of life: arthropods (which include spiders, insects, and lobsters and their relatives), annelids (earthworms and their like), corals, sponges, mollusks (including clams, snails, and squid), and echinoderms (sea urchins, starfish, and sand dollars). The current geological era, the Cenozoic, is often called the Age of Mammals, which reflects a very biased perspective on life. A more accurate judgment of success in life's game (in numbers at least) would make it the Age of Arthropods. Now, as for much of Earth history, arthropods dominate; they constitute some 40 percent of living species.

As I've noted, more species of animal exist today than at any time in Earth history, each a unique variant on one of those thirty fundamental body plans. This rise in biodiversity through evolutionary time is one of the mysteries of life on Earth, part of an inexorable pressure from that mainspring of evolution, generating myriad variations on a few basic themes. All these thirty-odd modern body plans trace their origin back to the Cambrian, back to that orgy of innovation that took place in the interval of 530 to 525 million years ago. If a potential phylum was absent in the Cambrian, it was doomed to be absent for all time. It was as if the facility for making evolutionary leaps that produced major functional novelties—the basis of new phyla—had somehow been lost when the Cambrian period came to an end. It was as if the mainspring of evolution had lost some of its power.

The Cambrian explosion should therefore be recognized as special not only in the intensity of innovation over so short a period of time, but also in its remaining unmatched, or even approached, in fecundity. Apparently there was something extremely special—perhaps unique—in the macroevolutionary mechanisms that operated just a little more than half a billion years ago. Jeffrey Levinton, a biologist at the State University of New York, Stony Brook, graphically formed the conundrum biologists face with this fact: "Why haven't new animal body plans continued to crawl out of the evolutionary cauldron during the past hundreds of millions of years?"[4] Two major answers have been formulated over the years. The first appeals to principles of ecology, the second to genetics, and both are highly theoretical. There are others, too, that are even more speculative.

Ecologically speaking, Earth in the early Cambrian was a simple place. There were hundreds of thousands of species of single-celled organisms and remnant populations of the Ediacaran fauna; but the niches occupied by the multicellular organisms in today's world were yet to be filled. The early Cambrian world therefore have offered unbounded ecological opportunity. "Ecosystems had room for everything—crawlers, walkers, burrowers, slurpers, predators, you name it—and life responded with unparalleled seizure of the opportunity," observed Gould. "Once exploited in the Cambrian explosion, such an opportunity could never arise again because the [available ecological space] never became so empty."[5] The empty ecological niche idea, which James Valentine of Berkeley advanced two decades ago, is within the standard Darwinian framework in that it looks to external influences to shape the world.

According to this idea, evolutionary innovation has remained constant through the past half billion years, but the probability of survival of its products has changed. In early Cambrian times, innovations at the phylum level survived because they faced little competition. Later, when all available niches had become occupied, further major innovation at this level failed, simply because there was no new ecological space to be exploited. The hypothesis is attractive, not least because it fits into conventional biological principles—always a good proposition. It has a pleasing simplicity and familiarity to it.

In 1987 Valentine put the hypothesis to the test, at least in the theoretical realm. To do this he joined forces with Douglas Erwin, now at the Smithsonian Institution in Washington, and Jack Sepkoski, at the University of Chicago, and scrutinized the fossil record. They noted that another remarkable event in Earth history—the biggest of all mass extinctions—which occurred at the end of the Permian period, 225 million years ago, wiped out as many as 96 percent of all marine species. This gigantic global catastrophe, known as the Permian extinction, offered an experiment of nature with which to compare what happened at two different periods in Earth history when the planet was ecologically impoverished. If ecological space was similarly vacant at these two periods, a similar response—a burst of evolutionary innovation—would be expected, according to the hypothesis.

"It appears quite unlikely that more than a few hundred to a thousand [multicellular] species coexisted in the marine biosphere at those times," observed Valentine and his colleagues. "In each case, a strong diversification ensued."[6] The relics of the fossil record reveal that the bursts of evolutionary innovation that followed the Permian extinction closely matched the innovation after the Cambrian explosion—but it was a quantitative, not a qualitative, match.

Biologists arrange living organisms in a hierarchy of classification, with species the lowest level and phylum virtually the highest (kingdom is on top); in between lie genus, family, order, and class. In the middle range of this hierarchy, the two periods of innovation matched closely. For instance, during the Cambrian, a total of some 470 new families appeared, which compares with 450 for the post-Permian. Below the level of the family, the Cambrian explosion produced relatively few species, whereas in the post-Permian a tremendous species diversity burgeoned. Above family level, however, the post-Permian radiation faltered, with few new classes and no new phyla being generated. Evidently, the mainspring of evolution operated in both periods, but it propelled greater extreme experimentation in the Cambrian than in the post-Permian, and greater variations on existing themes in the post-Permian.

At first blush, this might seem to invalidate the empty ecological niche hypothesis, because opportunities for the survival of major evolutionary innovations surely existed after the Permian crisis, and yet none occurred. Not so, argue Valentine and his colleagues; the hypothesis remains intact. True, 96 percent of marine species vanished in the Permian extinction, but the rest represented the complete spectrum of niches, even though their absolute numbers had been drastically reduced. It was as if, thinly spread though they may have been, the postcrisis creatures were still sufficiently diverse in their body plans to stake their claims across the entire ecological map. There was simply no room for major new body forms to endure, no unoccupied niche for radically different newcomers to become established. Variations on existing themes were possible, yes, but no major departures from them.

I can see why the hypothesis has its supporters; it makes good, intuitive sense. But the genetic hypothesis has to be reckoned with, too. Contrary to the ecological hypothesis, this alternative model suggests that the mainspring of evolution was qualitatively different in the early Cambrian from subsequent times. Specifically, it posits a more prolific production of major evolutionary innovation early in the history of multicellular life. The proposed reason for the difference is a "more loosely" organized genetic package, or genome, early on. As time passed, the different elements of the genome became ever more integrated, ever more refined, so that perturbations of the system were tolerated less and less. Bigger genetic changes—resulting in bigger morphological changes—are more feasible in a loosely organized genome than in one that is tightly integrated. Hence, evolution in Cambrian organisms could take bigger leaps, including phylum-level leaps, while later on it would be more constrained, making only modest jumps, up to the class level. That, at least, is what the hypothesis argues.

It's not easy to imagine how this hypothesis can be tested, as primordial genomes no longer exist for scrutiny in the laboratory. It is, however, plausible, as genomes are likely to have evolved in complexity, just as organisms have. There was an unusually fast turnover of species during the early Cambrian; that is, species originated and became extinct rapidly, compared with other periods in Earth history. Does this imply that evolution in these ancient times was less constrained because of loosely organized genomes? Perhaps. But an ecological explanation could be offered, too, with the notion that something external, some unusual condition in the environment, was driving the change. This position is supported by the observation that rapid species turnover occurred not only in multicellular species but also in single-celled organisms. Although several hypotheses have invoked environmental change—such as a shift in the chemistry of sea water that allowed calcified skeletons to form—convincing evidence for a physical driver of the Cambrian explosion remains elusive. Valentine recently listed twenty such hypotheses that have been proposed by various authorities over the years. So long a list surely betrays desperation for a solution among biologists as a whole rather than true insight into the problem by any one of them.

• • • •

Life's salient mystery therefore continues to baffle biologists—in its timing, its character, and its magnitude. But I have an admission to make: I have been holding back half the story, arguably the more fascinating half. The fact that the mainspring of evolution in the early Cambrian operated as I've described—that is, in generating all modern body plans in a split second of evolutionary time—is sufficient in itself to command our attention and our awe. In recent years, however, it has become obvious that the character of innovation was even more extravagant than anyone had imagined. And the history of life following the explosion was unimaginable. The implications for how we view Earth history are profound.

I said earlier that in that brief moment of creativity half a billion years ago, all modern phyla arose. In fact, the thirty or so animal phyla that form the modern biological world were initially in the company of perhaps as many as seventy others that have long since vanished. The Harvard biologist Edward Wilson described the time as "a period of wild experimentation during which basic body plans never seen before or afterward were invented and discarded."[7] To look back at that time by delving into the fossil record is to visit an alien world, populated by creatures built in ways we have never experienced. The dinosaurs seem wondrous enough to us, with their gigantic size and often fearsome weaponry, but they are familiar to the eye of a biologist, who sees them as interesting variations on terrestrial tetrapods, from crocodiles to elephants, from tree shrews to lions. Back in the early Cambrian, there were creatures that, albeit small in size, seem like something out of science fiction.

And yet, when the products of the Cambrian explosion were first seen and studied nine decades ago, they were viewed in a very different way. The story of this shift in perception is now a classic one in science, and it has been chronicled splendidly by Gould in his 1989 book, *Wonderful Life*. It is the story of two men and their devotion to what they separately saw as the truth.

In November 1909 Charles Doolittle Walcott discovered what Gould has called "the Holy Grail of paleontology"; namely, the Burgess Shale. Prospecting for fossils along the western slopes between Wapta Mountain and Mount Field in the southern part of British Columbia, Canada, Walcott happened upon what essentially was a marine Pompeii, or rather several Pompeiis compressed together like the leaves of a book. Repeatedly, shallow-water communities were entombed by sudden mudslides, burying and ultimately preserving glimpses of life a little more than half a billion years ago, soon after the Cambrian explosion. In a letter to a colleague at the time, Walcott noted that he had found "some very interesting things."[8] He was right, but for a variety of reasons he did not recognize their full import.

Periodically during the next two decades of his busy life, Walcott studied the massive amount of fossil material he and his team lugged back to Washington, D.C. He drew carefully what he saw, and assigned each creature to an existing phylum. In other words, Walcott saw in the Burgess Shale an image of today's world, but in more primitive form. Like Darwin, he fully expected that further fossil discoveries would furnish appropriate ancestors of the Cambrian fauna, in a long evolutionary chain. And, again like Darwin, he envisaged life progressing in a gradual fashion, with no dramatic booms or busts. As Walcott saw it, the exquisitely preserved Burgess Shale fauna "essentially set the evolutionary scene for the rest of Earth history," as I said in the opening paragraph.

The second phase of the story began in the late 1960s, when Harry Whittington, a paleontologist at Cambridge University, re-opened Walcott's excavation. "We camped at 7000 feet, and then climbed another 500 feet onto the ridge," Whittington recalled. "You are perched high above a sparkling green lake, and you look west toward the snow-clad Rockies. Best of all, you find the most wonderful fossils when splitting the rock."[9] Those fossils included sponges, jellyfish, worms, molluscs, and many arthropods. Some were bottom dwellers, either fixed or mobile; some were swimmers or floaters. For the first time, creatures with calcified skeletons were present in the unfolding pages of life. About 20 percent of the 140 or so Burgess Shale species were skeletonized, even though they represented only 5 percent in terms of numbers of individuals. No matter their mode of life, all members of the Burgess bestiary were preserved as flattened, ghostly images in thin layers of shale, their external anatomy intact in virtually every detail.

During the following decade Whittington recruited a small team of bright young researchers to help him study the wealth of material from the original and new excavations. Very soon, a pattern of discovery began to emerge: not only was it possible to identify creatures with modern descendants, just as Walcott had, but it was also clear that many of the Burgess Shale fauna had no descendants, contrary to Walcott's conclusion. For instance, Whittington and his colleagues could find representatives of the three types of living arthropod (spiders, insects, and lobsters and their relatives), and the favorite of all young fossil hunters, trilobites, a fourth form of arthropod that became extinct at the end of the Permian, 250 million years ago. But there were many other types of arthropod, none of which survived beyond the Cambrian.

The more Whittington and his colleagues scrutinized the ancient fauna, the more those creatures looked foreign to them. Eventually, a list of about twenty species "defied all efforts to link them with known phyla,"[10] recalled Whittington and Simon Conway Morris, one of his young colleagues. Known for good reason as *Problematica*, these bizarre creatures were built on no familiar body plan. They had burst onto the scene explosively in that brief period of "wild experimentation," and had disappeared just as dramatically. Whittington and his colleagues were able to recognize this entirely unpredicted pattern because they had not blinkered themselves with a narrow Darwinian outlook, as Walcott had. Walcott had assumed that Cambrian life was but a stage in the progression toward modern life, and he made links between ancient and modern where none existed. As so often happens in science, Walcott saw in the evidence before him what he wanted to see.

Whittington and his colleagues' discovery that the Cambrian explosion had generated a riot of new body plans, the majority of which were rapidly lost, begged a major issue: What determined which would be the winners and which the losers? "We might ask two related questions," noted Gould. "First, were the unique Burgess phyla doomed by inadequate design to their brief existence as failed experiments in the first flowering of animal life? Second, for living groups with Burgess representatives, would we have known, at this outset, which were destined for domination and which for peripheral status in the nooks and crannies of an unforgiving world?"[11] These questions go to the heart of how we view the history of life.

According to the traditional view—that is, Darwin's derived expression that "all nature is at war, one organism with another"[12]—competition is the arena in which anatomical design, honed by natural selection, is tested: superior design leads to victory in life's race; inferior design leads to oblivion. In their early work, this is precisely how Conway Morris and Whittington characterized events, in an article published in 1979. "[M]any Cambrian animals seem to be pioneering experiments by various [multicellular] groups, destined to be supplanted in due course by organisms that are better adapted," they observed. "The trend after the Cambrian radiation appears to be the success and enrichment in the numbers of species of a relatively few groups at the expense of the extinction of many other groups."[13]

The first step in the revolution surrounding the reinterpretation of the Burgess Shale, therefore, was to accept the notion that it reveals a world of evanescent existence: many more body forms evolved than eventually survived. The second step was to characterize the victims in this mass extinction as having succumbed through inferior design, in true Darwinian fashion. Not surprisingly, this became the canonical view. There was a third step, however, again unpredicted.

The more Whittington and his colleagues studied the Cambrian animals, the more they realized they could see no obvious competitive advantage in those which survived nor feebleness in those which disappeared. In *Wonderful Life,* Gould traced in Whittington and his colleagues' publications the gradual realization that perhaps competition wasn't the answer. "They talked less and less about 'primitive' designs, and labored more and more to identify the functional specializations of Burgess animals," observed Gould. "They wrote less about predictable, ill-adapted losers, and began to acknowledge that we do not know why *Sanctacaris* is cousin to a major living group, while *Opabinia* is a memory frozen in stone."[14]

If survival and extinction are not determined by which species are better adapted and which worse, then what is the explanation? "We must entertain the strong suspicion the Burgess decimation worked more as a grand-scale lottery than a race with victory to the swift and powerful,"[15] suggested Gould. Conway Morris clearly concurred, writing in a major article in *Science* in October 1989 that "[the] macroevolutionary patterns that set

The Cone of Increasing Diversity

Decimation and Diversification

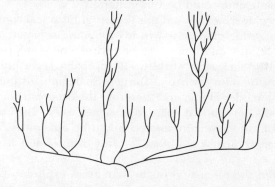

Two views of increasing diversity of body forms through time. Top: The traditional view that diversity increased gradually through time. Bottom: The notion that diversity was greater at the Cambrian explosion, after which many forms were lost. (Copyright Stephen Jay Gould/W.W. Norton)

the seal on Phanerozoic life are contingent on *random* extinction" (emphasis added).[16] Gone is the traditional notion of success through superior design, yielding a world populated by species of steadily improving adaptation and complexity. Enter a world populated by lucky survivors in a game of chance.

When I give lectures on human origins and on the complexity of ecosystems that many of us are striving to conserve, I listen to people's questions and hear a discomfort at the notion that anything in the history of life is merely random. I hear a yearning for an assurance that the universe of things is the way it should be, with our place in it predictable. I'll return to this issue in a later chapter. But here, in our overview of the beginning of complex life, it is proper that "we entertain the strong suspicion" that an element of randomness guides the history of life.

So when I said earlier that with the Cambrian explosion "the evolutionary scene was essentially set for the rest of Earth history," I was speaking only part of the truth. It is true that all extant phyla came into being then. But if randomness plays any significant role in the unfolding of Earth history and the survival of that initial burgeoning of life was not dependent on good design, then a whole different set of players might have occupied today's world. "The divine tape player holds a million scenarios,"[17] is the way Gould put it. Play the tape again, he suggested, and the outcome could be very different.

Conway Morris expressed this notion of historical contingency in the following graphic way in his 1989 *Science* paper: "What if the Cambrian explosion was to be rerun?" he pondered. "At a distance the metazoan world would probably seem little different; even the most bizarre of Burgess Shale animals pursue recognizable modes of life, and therefore the occupants of the ecological theater should play the same roles. But on close inspection the players themselves might be unfamiliar." The result, he said, might be a biota "worthy of the finest science fiction."[18]

I started the discussion of the Cambrian explosion with stromatolites on the western shore of Lake Turkana and finish with potential science fiction. It's been a giddy journey, but an essential one if we are to appreciate our place in the universe of things and our role in its future. I applaud Gould's advocacy of historical contingency as a major player in shaping the unfolding of life on Earth, so that we see our world as just one of many possible biological worlds. Gould does not deny that natural selection is important in adapting species to their environments, but sees it as a local influence, not one that shapes the broader history of life. This resonates with my own experience of the fossil record, but, like many evolutionary biologists, I believe Gould goes too far. For instance, I concur with Jeffrey Levinton's reply to the question *Design or accident?*: "I think the best to be said for now is that there is some truth in both alternatives."[19]

Earth history is surely shaped to a degree by random forces, particularly in selecting victims of mass extinctions. But it is intuitively obvious that life has progressed from the simple to the complex, both in anatomy and behavior. Where once the world was populated with organisms no bigger than a single cell, there are now myriad species constructed from many different types of cells. Where once the world was populated with organisms that merely responded to their environment as automatons, there are now myriad species that reflect before they act. Where once the world was without a single species that possessed a sense of self-awareness, there is now at least one species so blessed. The mainspring of evolution has been productive indeed.

4

The Big Five

I'M OFTEN ASKED how my colleagues and I find fossils. The answer is more simple than most people expect: we simply walk around and look for them. You have to be in the right place, of course; that is, looking at rocks that were formed millions of years ago and are currently being exposed by erosion. And the laying down of the sediments that formed the rock had to have been under circumstances that favor the preservation of bones, such as in environments near rivers and lakes. Most of the sediments around Lake Turkana, in northern Kenya, where I've been finding ancient human fossils since 1969, were laid down between one and four million years ago. Wind and rain and seasonal streams effectively leaf back through the pages of geological time as they erode ancient sediment and expose the bones of our ancestors who lived there long ago.

In addition to relics of early humans, at Lake Turkana we find petrified bones of pigs, crocodiles, elephants, monkeys, antelopes of various kinds, and many other species. With a little imagination, it is possible to conjure up in one's mind a sketch of that ancient life, and it would be very similar to life there today. There would, however, be a key distinction: all the species in the imagined ancient scene would be different in some way from the extant community of species, some slightly so, others dramatically so. There was a crocodile with an extraordinarily long snout, for instance, an elephant whose tusks dropped down from the lower jaw, unlike the upward sweep of today's species, and a very small species of hippo. Moreover, there were many more species then, including three species of elephant, three hippos, and several pigs. So when I walk across the sandstone reaches around Lake Turkana and see relics of creatures that lived two, three, or four million years ago, I'm seeing more than the remains of ancient ecological communities; I'm seeing evidence of communities that no longer exist because the species no longer exist. They are extinct.

To a paleontologist, death is a fact of life, and extinction is a fact of evolution. Some thirty billion species are estimated to have lived since multicellular creatures first evolved, in the Cambrian explosion. According to some estimates, thirty million species populate today's Earth. This means that 99.9 percent of all species that have ever lived are extinct. As one statistics wag put it, "To a first approximation, *all* species are extinct." Obviously, not all; otherwise we wouldn't be here to contemplate the wonder of nature. But life's grip on Earth is evidently more precarious than we might like to accept.

In the 530 million years since the Cambrian explosion, thirty billion species have evolved, some of which were slight variants on existing themes while others heralded major adaptive innovations, like jaws, the amniote egg, and the capacity of flight. Quite properly, biologists have been riveted by the mechanisms that give rise to such innovations, and specifically by the processes that produce new species. The study of evolutionary biology and ecology is principally about speciation and the interaction of species within communities. The average life span of animal species is four million years, so during the same 530 million years only a tiny fraction less than thirty billion species have gone extinct. It seems intuitively obvious that the balance between speciation and extinction would be important in the history of life on Earth, and yet the subject of extinction has all but been neglected by evolutionary biologists until very recently.

There are several reasons for this neglect, although two factors conspired to turn it around. The first was the suggestion, made a little more than a decade ago, that the dinosaurs met their end when a huge asteroid collided with the Earth, sixty-five million years ago; the second was the growing realization that we are witnessing modern extinctions on a cataclysmic scale, the product of human encroachment on ecosystems. The first caught the imagination because of its inescapable drama; the second through awakening a concern about our tenure here on Earth. As a result of the recent burst of research into this most ancient of processes, biologists' assumptions about extinction—about its causes and, more important, its effects—have been overturned.

This chapter will discuss two aspects of extinction, or rather mass extinction, periods in Earth history when a significant fraction of the existing species was extirpated. First is the *fact* of such events, something that Darwin and others vigorously denied. Second is the *cause* of these mass dyings, a subject of much debate and little agreement.

Two further aspects of mass extinction, discussed in the following chapter, are, first, the effect on the history of life and, more specifically, the question of what determines which species survive these extinctions and which succumb. The second is their aftermath; that is, the biota's response to decimation. With much uncertainty surrounding these several issues, one thing has become clear in the past few years: once considered a passive component in the course of evolution, mass extinction is now recognized as a major determinant of its outcome.

As with many issues in evolutionary biology, Darwin's influence on modern thought concerning extinction has been profound. In *The Origin of Species,* Darwin noted the existence of extinction, saying, "No one I think can have marvelled more at the extinction of species, than I have done."[1] He recognized four key features in the phenomenon. First, that it was a continuous and gradual process: ". . . species and groups of species gradually disappear, one after another, first from one spot, then from another, and finally from the world."[2] Second, that the rate of extinction was essentially uniform and did not occasionally surge, producing mass extinctions, as many paleontologists of his time believed: "The old notion of all the inhabitants of the Earth having been swept away at successive periods by catastrophes, is very generally given up."[3]

Third, that species go extinct because they are in some manner inferior to their competitors: "The theory of natural selection is grounded in the belief that each new variety, and ultimately each new species, is produced and maintained by having some advantage over those with which it comes into competition; and the consequent extinction of less favored forms inevitably follows."[4] Fourth, that extinction is an integral part of natural selection, as implied in the previous citation. He also wrote, "Thus the appearance of new forms and the disappearance of old forms . . . are bound together."[5]

Two issues here are central to current scientific debate: the notion that species go extinct through failure in competition; and the tempo of extinction through time.

Darwin's equating of extinction with adaptive inferiority clearly derives from his theory of natural selection, and it has, until very recently, powerfully shaped biologists' thinking. Many paleontologists of Darwin's time argued that the fossil record indicated occasional bursts of extinction, crises in the history of life. The evidence for such mass extinctions must be due, said Darwin, to the incompleteness of the fossil record—an artifact of the record, not a reflection of reality. His insistence that extinction proceeded gradually and without occasional sharp increases stems from two factors. First, natural selection is a gradual process, so extinction must be gradual too, because they are linked. The second argument reflected the newly emerging paradigm in geology, that of uniformitarianism. It's worth pursuing this briefly, because it illustrates the source of Darwin's bias in this respect as well as a similar bias among many modern paleontologists, manifested in their response to the suggestion that asteroid impact caused the demise of the dinosaurs.

The fact of extinction was established by the French anatomist Baron Georges Cuvier, in the late eighteenth century, when he demonstrated that mammoth bones are different from those of the modern elephant. The inescapable conclusion was that mammoth species no longer existed. Through his extensive study of fossil deposits in the Paris Basin, Cuvier went on to identify what he thought were times of crisis, or catastrophes, in Earth history, when large numbers of species went extinct in very short periods of time. His observations inspired a great volume of geological work in the early part of the nineteenth century. This work identified intervals of apparent major change, which formed boundaries between geological periods that were given the following names: Cambrian, Ordovician, Silurian, Carboniferous, Permian, Triassic, Jurassic, Cretaceous, Paleocene, Eocene, Oligocene, Miocene, Pliocene, Pleistocene, Holocene. The bane of geology students, who have been more than inventive in devising interesting mnemonics for remembering them, these names remain part of the geological time scale used today.

Cuvier pointed to what he considered two particularly devastating catastrophes that divided the history of multicellular life, known as the Phanerozoic, or visible life, into three eras: the Paleozoic (ancient life), from 530 to 225 million years ago, the Mesozoic (middle life), from 225 to 65 million years ago, and the Cenozoic (modern life), from 65 million years ago to the present. Conversations with geologists often involve mind-boggling blizzards of the names of eras and periods, because to them, as to Cuvier, they represent a true record of the history of life. Cuvier lived in pre-evolutionary theory times, of course, and therefore saw the catastrophes as individual events that wiped out existing life, setting the stage for new waves of creation. The Noachim Flood was said to have been one such event; the total number of crises was eventually estimated to have been about thirty. Cuvier's scheme came to be known as catastrophism.

Even before Darwinian theory came along, catastrophism was under attack, a move spearheaded by the Scottish geologist Charles Lyell, who was following a path blazed earlier by his countryman James Hutton. In the early 1830s, Lyell published his three-volume *Principles of Geology,* in which he argued that the geological processes we observe today—such as erosion by wind and rain, earthquakes and volcanoes, and so on—are responsible for all the geological changes throughout Earth history. He also denied the existence of mass extinctions of species. "[We] are not authorized . . . to recur to extraordinary agents" to explain apparently large changes in the geological past, he asserted. Over long periods of time, he argued, small changes accumulate to produce large effects. Lyell's dictum became embodied in the phrase "The present is the key to the past," meaning that we can use our own experience to understand events and processes of

All societies have had (prescientific) mythologies about how the Earth was formed. Here we see Pan-Kou-Ché, the Chinese creator, a feeble old man who toils painfully at his work, covered with perspiration, sculpting the Earth's crust from a confused mass of rock.

the past. This, however, is logically flawed. Human experience of what is geologically possible is extremely limited, because we have been here for so short a time, and it is highly unlikely that all possible geological processes that can shape the Earth have occurred for us to witness.

Lyell's scheme came to be known as uniformitarianism, and for a while an intellectual battle was joined between it and catastrophism. Uniformitarianism won decisively, and catastrophism was banished from the intellectual arena as a relic of earlier thinking, governed, it was implied, by religious rather than scientific consideration.

For Darwin, the ascendancy of uniformitarianism in geology provided the foundation for his own gradualist theory in biology: namely, natural selection, in which small changes accumulate over a long period of time, resulting in large evolutionary change. And just as Lyell had rejected the notion of catastrophes in geological history as an appeal to "extraordinary agents," so Darwin rejected the notion of crises in the history of life as an appeal to supernatural intervention, that of catastrophism. He had struggled long to replace a supernatural explanation of life's exquisite adaptation to its circumstances by a naturalistic one. He therefore balked at anything that smacked of the supernatural, such as the sudden extirpation of millions of species followed by waves of creation.

Catastrophism may have been defeated in academia, but the catastrophists' notion of the pattern of life in Earth history remained stubbornly persistent. The more that geologists and paleontologists clambered over ancient exposures, seeing remnants of mass dyings, the more certain they became that from time to time catastrophic events had indeed occurred. Even as the record became more complete, with the gaps much lamented by Darwin filled in here and there, the evidence of periodic crises became ever more compelling. Earth history evidently is not one of gradualistic progression, as Lyell and Darwin fervently desired, but one of sporadic and spasmodic convulsions. Some of these were of moderate extent, in which 15 to 40 percent of marine animal species disappeared, but a few others were much larger.

This last group—known as the Big Five—comprises biotic crises in which at least 65 percent of such species became extinct in a brief geological instant. In one of them, which brought the Permian period and the Paleozoic era to a close, it is calculated that more than 95 percent of marine animal species vanished. As the Chicago University geologist David Raup remarked of the Permian extinction, "If these estimates are even reasonably accurate, global biology (for higher organisms at least) had an extremely close brush with total destruction."[6]

Many of the big changes in Earth history are inferred from the marine fossil record, for the very good reason that it is so much more complete than the terrestrial record. (The bones become covered in silt faster, and thus begin their journey into the fossil record. On land, carcasses are ravaged by scavengers, trampled by passing herds, and may lie in the hot, dry air for centuries, crumbling into dust and never becoming entombed intact in sediments.) Even though there were in the past probably ten to a hundred times as many animal species living on land than in the sea, as is the case today, some 95 percent of the 250,000 species known in the fossil record are marine animals. This illustrates how poor a picture of past terrestrial life we have. For an extinction burst to qualify as a mass extinction, however, paleontologists demand that the catastrophic effect be evident in both the marine and terrestrial records, for this betrays a global event.

If we look at a visual record of the diversity of life on Earth—that is, the number of species that exist at any one time—we see that it begins very low in the Cambrian and rises to a high point in the present. Darwin thought that the increase followed a steady course, but that was not the case. The Big Five interrupted that rise dramatically, periodically plunging diversity to dangerously low levels. So, graphically, instead of a straight line sloping up from left to right (following a conventional scale of time), we see a jagged, saw-tooth pattern. This handful of major events, from oldest to most recent, are: the end-Ordovician (440 million years ago), the Late Devonian (365 million years ago), the end-Permian (225 million years ago), the end-Triassic (210 million years ago), and the end-Cretaceous (65 million years ago).

Paleontologists realized from the beginning that the biotic crises were more than an interruption in life's flow, however, for at each event the character of ecological communities shifted, sometimes dramatically so. (This was what Cuvier had interpreted as the outcome of new waves of creation.) The most famous of such shifts, of course, was the end-Cretaceous event, sixty-five million years ago, which saw an end to 140 million years of terrestrial domination by dinosaurs. In the subsequent Cenozoic era, mammals came to dominate vertebrate life on land. Similarly, the great Permian extinction dealt a near fatal blow to the mammal-like reptiles, which had ruled terrestrial life for eighty million years. They never recovered, and their role was soon taken by the dinosaurs. Each mass extinction displayed the

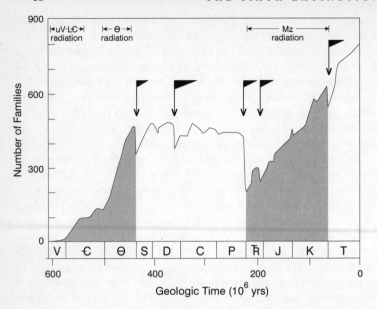

Five times in the history of life mass extinctions have dramatically reduced stand-ing biological diversity. Notice how diversity has increased through time, despite these occasional crises, reaching a maximum near the present. (Reprinted by permission of David Raup and Jack Sepkoski)

same pattern, but this is not the place to go into detail of the changes in the terrestrial and marine realms. Paleontological texts should be sought for that. The *pattern* of the extinctions is what is important in our discussion.

If the history of life is viewed as a drama staged on planet Earth, then it can be seen having repeated intermissions, after each of which the cast on stage changes: some characters, previ-ously important, disappear entirely or assume minor roles; oth-ers, in the wings, now move to stage front in major roles; new characters sometimes appear, too, producing a constantly shift-ing, Alice-in-Wonderland effect. Inevitably, fundamental shifts in *dramatis personae* force fundamental changes in the story line. So it is with Earth history. *Homo sapiens* is one of a set of characters whose presence on stage was influenced by the upheaval of the last mass extinction, the end-Cretaceous.

A question that has come into focus in recent years is: How

Full History of Life

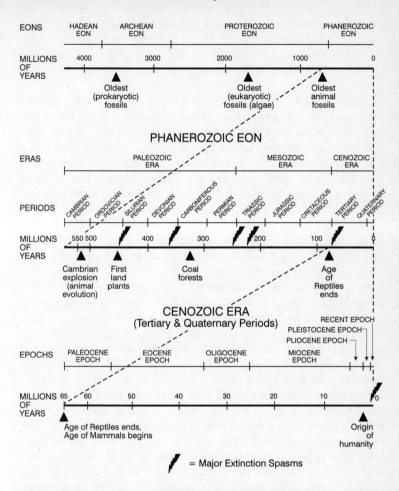

/ = Major Extinction Spasms

The full geological history of life goes back more than 3.5 billion years, when the first single-cell organisms appeared. Key episodes in evolution are placed within the divisions of geological time: eons divided into eras, eras into periods, and periods into epochs. Biodiversity was sharply reduced by the great extinction spasms, indicated here by lightning flashes.

(Reprinted by permission of the publishers from The Diversity of Life *by Edward O. Wilson, Cambridge, Mass.: The Belknap Press of Harvard University Press, Copyright © 1992 by Edward O. Wilson.)*

predictable is the shift in the cast of characters? In other words, what determines which species survive and which succumb in mass extinction events? I will explore this in the following chapter. Here, I will address the question of what triggers the crises in the first place. What is the cause of mass extinctions?

The satirist Will Cuppy had an answer, or at least he did for the dinosaurs. In his 1941 book, *How to Become Extinct,* he wrote, "The Age of Reptiles ended because it had gone on long enough and it was all a mistake in the first place." Cuppy may have had a point, and, in any case, evolutionary biologists and paleontologists have long shied away from trying to explain the cause of mass extinctions, because they were considered too complex to understand.

There has been no shortage of suggested causative agents of mass extinction events over the decades, some of them wildly unlikely, such as the lethal effects of nearby exploding supernovas. Among the hypotheses that have been taken seriously, however, are global cooling, falling sea levels, predation, and competition among species. Arguments can be adduced in support of each of these, and each has its enthusiastic proponents. Collectively, however, they display a characteristic that goes beyond scientific plausibility, says David Raup. "I see more than a hint of anthropomorphism here," he wrote in his 1991 book, *Extinction: Bad Genes or Bad Luck?* "Could it be that the list of probable causes of extinction is simply a list of things that threaten us as individuals?"[7] It is surely true that extinction holds a certain horrific fascination for us, because if species or groups of species that have been successful for millions of years, such as the dinosaurs, can slip into evolutionary oblivion, what about *Homo sapiens?* Are we vulnerable too? Many of the attitudes that govern discussion on extinction reflect emotional as well as scientific viewpoints.

Those who have compared the global conditions that prevailed during each of the Big Five note that a common factor is a fall in sea level, or marine regression. Sea level may fall for any one of several reasons, including extensive polar glaciation and changes in the configuration of the continents as they ride the moving plates that form the Earth's crust. The potential effect on marine life in shallow waters is dramatic. As the sea level drops, continental shelves become exposed, thus reducing available habitat for shallow-water species. Biologists are well aware of the relationship between available habitat and species number, and it is simple: the less area available, the fewer the species that can exist. So marine regression does seem a good candidate for triggering a mass extinction. Except that, by definition, mass extinctions must affect not only life in shallow waters, but in the deep, too, and on land. There is a possible connection between falling seas and mass killings on land, however, as Paul Wignall, a paleontologist at the University of Leeds, England, has pointed out in connection with the end-Permian extinction.

This event, remember, extirpated more than 95 percent of marine animal species and almost as many on land. Whatever was going on at that time must have been remarkably efficacious. The end of the Permian period coincided with a time when all the continents of the world coalesced, through continental drift, forming a single supercontinent, Pangea, which stretched from pole to pole. This fact alone reduced the available habitat for shallow-water species. Imagine four one-inch squares, each of which has a total edge length of four inches, giving a grand total of sixteen inches. Now bring them together as a single square of side two inches. The total edge length is now a mere eight inches, just half of the previous figure. The same thing happens with individual continents and available shallow-water habitats. The formation of Pangea therefore must have devastated species in these habitats by this mechanism alone, as Stephen Jay Gould argued some years ago.

Combine this with marine regression, and disaster clearly awaited species in these vulnerable habitats. And, suggests Wignall, as the continental shelves lay exposed and dry, they underwent erosion and oxidation of organic matter that once was deep below the sea bottom. Extensive oxidation of organic matter sucks oxygen out of the atmosphere and pumps carbon back in return. Atmospheric oxygen may have dropped to half of today's level. Terrestrial animals, especially active vertebrates, would have been especially vulnerable to such an atmospheric shift. At the end of the Permian, sea levels rose rapidly once again, a process that not only reduced terrestrial habitats but also reduced oxygen in sea water, by some unknown mechanism. As a result, says Wignall, "the Permian-Triassic mass extinction appears to be a story of death by suffocation for both terrestrial and marine life."[8]

Douglas Erwin, of the Smithsonian Institution, agrees that more than one agent operated at this most massive of mass extinctions; it involved "a tangled web rather than a single mechanism."[9] He cites, in addition to marine regression and transgression, climatic instability caused by the configuration of the supercontinent and by carbon dioxide released by massive eruption of lava in Siberia, which formed an expanse of volcanic rock some 870 miles in diameter near Lake Baikal. I mention these explanations not because they have been proven to be true, but as examples of the level of complexity of events leading to the killing of species on a massive scale. One last comment about marine regression and its potential as a causative agent in mass extinctions: most, but not all, massive extinctions are associated with a fall in sea level; however, marine regression is not always associated with mass extinctions. Clearly, other factors must be involved.

The second most favored candidate as a messenger of mass death is global climate change, particularly global cooling. According to Steven Stanley, a paleontologist at The Johns Hopkins University, changing climate is *the* most important cause of crises in the history of life. Stanley, who has done extensive work on species extinction among shallow-water animals in North America, identifies climatic cooling as the major culprit. The mechanism is simple and direct, and has the merit of affecting both marine and terrestrial organisms. "There is one simple fact that makes climate change a likely general cause for mass extinction," observes Stanley. "This is the relative ease with which a change in global temperatures can eliminate myriads of species."[10]

Species are adapted to their local conditions, including food resources available to them and prevailing temperature, whether Arctic, temperate, or tropical. If the Earth cools, then habitats (as determined by temperature range) shrink toward the tropics. Global temperature has fluctuated substantially and rapidly throughout Earth history, and the typical response of species to such changes is migration: they follow the shifting climate, toward the equator in times of cooling, away in warming periods. Maps of forest cover in North and South America during the past twenty thousand years show this effect wonderfully, as the climate shifted out of an ice age and into an interglacial period, today's climate. At the beginning, the Amazonian forests were mere scattered fragments, and the deciduous and coniferous forests across North America had migrated south. When the ice receded, the forest refugia in the Amazon expanded and coalesced, and oak and larch began their northerly march in North America, following in the root steps of conifers.

Such migrations are never simple, with habitats moving *en masse;* instead, species scatter in different directions, later to form communities of different composition. Nevertheless, the phenomenon illustrates the need of species to move when climate changes. If they can. Sometimes the climate change is too rapid or too extensive for species to be able to respond; and geographical barriers such as rivers and mountains can block the only route available. When this happens, extinction is the most likely outcome. Extensive glaciation has occurred periodically in Earth history, sometimes, but not always, coinciding with mass extinctions. Therefore, although global cooling is undoubtedly important in many biotic crises, it cannot be the primary cause in all, as Stanley argues.

Marine regression and global climate change are just two of a panoply of potential proximate causes of extinction, but they stand out as the most important. In both cases, however, the proximate cause may be the result of different ultimate causes, such as changes in mantle convection, configuration of the tectonic plates of the Earth's crust, the variation in the Earth's orbit, and so on. All these mechanisms, both ultimate and proximate, are within human experience or recent history, and therefore are favored by those paleontologists who have bothered to think about mass extinctions at all. The uniformitarian paradigm ruled. It came as more than a little surprise, therefore, when, in 1979, a physicist, a geologist, and two chemists suggested that at least one mass extinction, the end-Cretaceous, was the result of an asteroid or comet colliding with the Earth. Dust debris was lifted into the high atmosphere, plunging the planet into virtual darkness long enough to kill plant life, on which animal species depended. Not surprisingly, as Gould later observed, paleontologists' reaction "ranged initially from skepticism to derision."[11]

• • • •

The story has been told many times, so I will be brief and will highlight the sociological as well as the scientific issues involved.

In the late 1970s, a team of scientists at the University of California, Berkeley, headed by the physicist Luis Alvarez, were using chemical methods to measure deposition rates in various sedimentary formations. To their surprise, while working in the Umbrian Apennines of Italy and in Denmark they found unusually high levels of a heavy, unreactive metal, iridium, in a thin clay layer that marks the end-Cretaceous mass extinction, sixty-five million years ago. Because it is heavy, iridium sank to the Earth's interior in the planet's early history, when much of the rock was molten; the metal is therefore rare in the Earth's crust and in continental rock. It is, however, a significant component of the minerals of meteorites. Alvarez and his colleagues put two and two together, and stunned the paleontological community: the Cretaceous event was caused by Earth's collision with an asteroid.

From measurements of the iridium present, the Berkeley team calculated that the asteroid was some seven miles in diameter. The energy from such a collision would have been immense, some billion times that of the Hiroshima atomic bomb. The impact would have created a crater about a hundred miles in diameter and injected sufficient debris into the atmosphere to blanket the sun and make perpetual night. Initially Alvarez and his colleagues estimated that the darkness would have lasted several years, but later revised their calculation to several months. This would still have been sufficient to devastate plant life on land and in the sea; animal life, dependent as it is ultimately on plants, would have followed.

The Alvarez announcement, published in *Science* in June 1980, sparked a flurry of conferences and scientific papers on the topic of asteroid impact. The tenor of discourse was not always calm, for the very good reason that "a large rock falling out of the sky is anathema to rank-and-file geologists," according to David Raup, who was to play a major role in the unfolding and extension of the story. These geologists, he said, "were taught that this sort of *deus ex machina* explanation for natural events should be avoided because it constitutes a return to the mysticism of the early days of science."[12]

You only have to aim a pair of binoculars at the moon to see that asteroid impacts occur in the solar system. The face of the moon is pockmarked with stark testimony to the fact. If the moon was bombarded, so too the Earth must have been. The evidence of terrestrial impact is, however, much less evident, partly because two thirds of the globe is ocean, so that craters formed there would be difficult to detect and, because of sea floor spreading, would eventually disappear into the bowels of the Earth. In addition, craters formed on land soon become subject to erosion, so that their initially crisp form slowly fades. Nevertheless, there are about a dozen large impact craters, such as the Manicougan crater in Quebec (eighty miles in diameter), the Siljan crater in Sweden (forty miles in diameter), and the Popigai crater in the former USSR. There is therefore no denying the fact of impact.

Several items of evidence had to be amassed, however, before the notion that impact can cause a mass extinction could be taken at all seriously, even by those who were sympathetic. First, high levels of iridium had to be found wherever sediments of the end-Cretaceous boundary were exposed. They were; and by now more than a hundred such sites have been tracked down. Second, other clues to massive impact should have been evident. They were, in the form of tiny glasslike spherules, which form from rock minerals under enormous temperature and pressure. These have been detected in seventy-one sites around the globe. Another clue to massive impact is so-called shocked quartz, crystals showing lines of shock fractures that result from sudden pressure. By now, shocked quartz has been detected at about thirty sites world wide.

The smoking gun, of course, would be evidence of asteroid impact at the Cretaceous-Tertiary boundary. As I indicated earlier, there is a high probability that the end-Cretaceous impact was in the ocean, which reduces the chances of our finding evidence. On land, one candidate is the Manson structure in Iowa. It is of the right age, but is probably too small, measuring only twenty miles in diameter. Paleontologists and geologists identified several moderately big craters of the right age, some in North America, some in Asia. Then, in June 1990, exactly a decade after the original Alvarez announcement, news spread in the geological community of the discovery of a huge crater underlying the northwest tip of the Yucatán Peninsula, Mexico. Named the Chicxulub crater, the structure was dated two years later by a team of geochronologists from Berkeley, led by Garniss Curtis, who, collaborating with my father, Louis Leakey, in the early 1960s, pioneered technical dating of early human fossils. The formation of the Chicxulub crater was put at sixty-five million years ago, precisely the time of the end-Cretaceous extinction. This, together with smaller craters of similar age, implies that perhaps a shower of asteroids, or asteroid fragments, bombarded Earth at about the same time.

The report, in *Science,* of the dating of the Chicxulub crater drew support, even from some previous skeptics of the impact theory. For instance, William Clemens, also at Berkeley, said in a commentary accompanying the paper, "Clearly, we've been underestimating cratering rates . . . Impacts are part of the environment of the [past 600 million years]."[13] Nevertheless, in common with many previous skeptics, Clemens, in accepting the idea of an end-Cretaceous impact, did not necessarily accept that it alone was the cause of the extinction. In the same commentary, Clemens cited Anthony Hallam, a paleontologist at the University of Birmingham in England, as saying: "I may accept the story of impact, but I think it was at most a coup de grâce. I believe a mass extinction would have taken place in the marine realm even without an extinction."[14]

Acceptance of huge impact by those who previously rejected the notion was progress of sorts, for it dented strict Lyellian uniformitarianism. It was a great philosophical breakthrough for geologists to accept catastrophe as a normal part of Earth history. From the very beginning of the debate, however, a vocal group of paleontologists and geologists argued that every one of the so-called clues to impact—high iridium levels, spherules, and shocked quartz—can result from massive volcanic activity. Volcanic ash is sometimes found at the end-Cretaceous boundary. And there are examples of huge outpourings of basaltic lava (flood basalt) from deep within the Earth that occurred sixty-five million years ago, such as the Deccan Traps in India, a series of plateaulike giant steps. It has also been suggested recently that massive impact and massive volcanic activity might be linked: energy from such an impact could be transmitted through the Earth's molten interior, triggering volcanism on the opposite side of the globe. Such a causal link may explain the Chicxulub impact and the Deccan Traps volcanism.

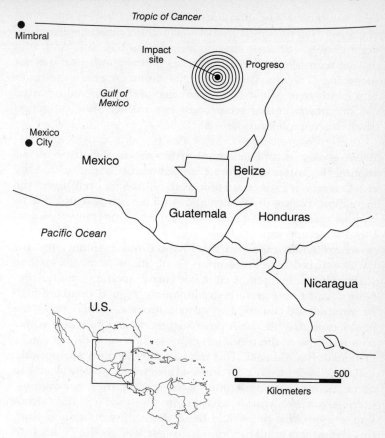

Evidence of a vast crater on the Yucatán Peninsula, some 65 million years old, is the smoking gun of the cause of the end-Cretaceous extinction: asteroid impact.

Although the debate lingers on, the chances look strong that asteroid or comet impact either caused or contributed significantly to the mass dyings at the end of the Cretaceous period. The notion that impact may have been only a *coup de grâce* is based on the suggestion that life was already in trouble when the asteroid hit. The initial claim by Alvarez and his colleagues met a counterclaim that the dinosaurs and ammonoids—stars of the terrestrial and marine tragedy respectively—were already in severe decline. Recently, this has been shown to be an artifact of the fossil record. The dinosaurs and ammonoids were doing fine at impact. There is evidence of decline among some groups of organisms, particularly marine organisms, however. And as a marine regression was in progress at the time, perhaps this is not surprising. The argument that major biotic crises are the result of a confluence of detrimental effects, perhaps including asteroid or comet impact, seems very reasonable. Mass extinction surely is a complex process.

While geologists and paleontologists were in the throes of accepting that asteroid or comet impact might have triggered the end-Cretaceous extinction, they were offered an even bigger pill to swallow. Basing their calculations on a newly assembled database of marine fossil species, the Chicago University geologists David Raup and Jack Sepkoski suggested in 1984 that the twenty or so extinction events in the Phanerozoic, including the Big Five, occurred at regular intervals of about twenty-six million years. The only agent that could drive such regularity, they opined, must be extraterrestrial; namely, regular bombardment by asteroids or comets. For various reasons, comets make the better candidate. If the Alvarez suggestion was met with skepticism to derision, the reception of periodic bombardment can be imagined: flat disbelief. This was catastrophism on a large scale.

Raup and Sepkoski's contention flows from a statistical analysis of extinctions through time, showing that they bunch every twenty-six million years, sometimes massively so. (On this calculation, the next big one would be in about thirteen million years, so we can relax a little.) Opponents, particularly the Polish paleontologist Antoni Hallam, counter by saying that the pattern is an artifact of the construction of the geological time scale. The debate has been going back and forth like a Ping-Pong match, with no resolution. If Raup and Sepkoski are right, then tell-tale signs of impact should accompany all major extinctions. So far, iridium signatures have been found at seven of the twenty extinction events, including the end-Cretaceous. But no such signature has been found at the end-Permian, the biggest of them all. Half a dozen craters have been lined up in age with major extinction episodes.

Raup and Sepkoski remain convinced by the power of their statistical analysis. If they are right, then fully 60 percent of all

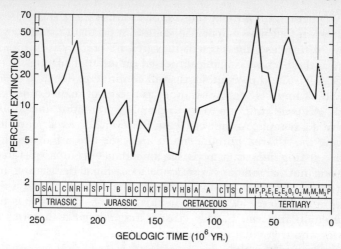

According to some views, significant extinctions occur regularly, every 26 million years, including the Big Five. A suggested cause is period bombardment of Earth by asteroids. (Reprinted by permission of David Raup and Jack Sepkoski)

extinctions through the Phanerozoic were caused by asteroid or comet impact (5 percent in the Big Five, the rest in the more numerous moderate events). Raup does not argue that impact alone causes the extinction event; he says that it is a "first strike," which weakens the biota, making it vulnerable to other inimical environmental processes that may be taking place. The notion that life on Earth is periodically assaulted by asteroid or comet impact seems like the stuff of science fiction. Nevertheless, the U.S. National Aeronautics and Space Administration is taking no chances. It is striving to set up a Spaceguard Survey, which comprises a system of Earth-based telescopes to watch for incoming bodies, with the backup of nuclear-tipped missiles that could be loosed to divert one headed for a direct hit.

Two recent near misses are reminders that the Big One is possible, one in March 1989, the second in January 1991. Both were small, measuring no more than three hundred yards in diameter, but both passed by at a distance comparable with that of the moon. And in July 1994, astronomers watched as a shower of some twenty-one comet fragments plunged into the surface of Jupiter, inflicting a series of huge scars on the planet's thick atmosphere, one of them the size of Earth. It was a timely demonstration that mass impact can and does occur in the solar system. In this case, the present may well be the key to the past, even if the present is being witnessed on another planet.

Geologists and paleontologists will continue to debate the feasibility of impacts versus volcanoes as agents of mass extinction; and NASA will continue to try to establish grandiose, and probably futile, systems. At the very least, however, the experience following the Alvarez announcement a decade and a half ago has ushered in a new catastrophism, one that encompasses forces outside normal human experience: occasional devastating impact from outer space. This represents the second major revolution in the science of geology in this century. The first was the realization that the Earth's crust is fragmented as a series of plates whose gradual movement through the eons moves continents around the globe.

5

Extinction: Bad Genes
or Bad Luck?

T HE HISTORY of life on Earth is punctuated by occasional
bursts of extinction, some moderate, some catastrophic; of
that there is now no doubt. Whatever the cause of these events,
the next crucial question we face is this: How does Earth's biota
respond? We've seen that major extinction episodes often leave a
different cast of players on life's stage, shifting the course of
evolution, sometimes dramatically so. What favored the survivors
at the expense of the losers? As David Raup so succinctly queried:
Was it bad genes or bad luck that consigned the losers to evolutionary oblivion?

For a very long time, this question did not need to be asked,
because the answer was an unspoken assumption of prevailing
evolutionary theory. In *The Origin of Species,* Darwin had been
very clear about why species went extinct: they were inferior to
their competitors. "The inhabitants of each successive period in
world's history have beaten their predecessors in the race for life,
and are, insofar, higher in the scale of nature," he wrote in his
chapter on the "Geological succession of organic beings."[1] The

heart of natural selection is the steady and continuous adaptation of species to their environments, an important component of which is other species, their competitors. Life is a continuous "struggle for existence."

In another section of the *Origin,* Darwin used his famous metaphor of wedges to describe this struggle: "The face of Nature may be compared to a yielding surface, with ten thousand sharp wedges packed close together and driven inwards by incessant blows, sometimes one wedge being struck, and then another with greater force."[2] When one wedge sinks in deeper, another or several others may be forced out. Nature is packed with species, each ultimately connected to the others through the force of competition, each struggling to survive. "The metaphor of the wedge underlies and supports our conventional view of life's order," observes Stephen Jay Gould, who has written extensively on this subject in recent years. "Creatures strive to improve themselves; life moves steadily upward although no one gets permanently ahead; order rules as the predictable struggle of individuals translates to patterns of increasing complexity and diversity."[3] The winners thrive, the losers become extinct, in the inexorable progress toward biological improvement.

The notion that extinction results from the failure of a species seemed so obvious that few considered it worth putting to the test. Tested it wasn't, for a very long time, as competition came to be accepted as the driving force of evolution and the differentiator between winners and losers. Each player in the game strives to outwit its competitors, through evolving enhanced behavior or anatomy; its opponents respond by evolving countermeasures. As a result, species change through time, improving in a sense, but rarely getting far enough ahead to overwhelm their adversaries. The University of Chicago paleontologist Leigh Van Valen named this effect the Red Queen hypothesis, after the character in *Through the Looking-Glass* who had to keep running merely to stay in the same place. If you fail to run, you fall behind and, in nature, become extinct. Such a fate would fall under the category of bad genes.

In recent years, biologists started to question the assumption of the ascendancy of competition, in local population interactions and at the level of evolution. As a result, competition is no longer perceived as the powerful force it once was. One instance of the hegemony of competition being undermined was work by Gould and his colleague C. Brad Calloway. Their subject of study was the relationship between brachiopods and clams, whose evolutionary roots go deep into the history of life. Clams dominate today's seas, while brachiopods play a minor role; millions of years ago, the reverse was true. This switch, explains Gould, has often been cited as a classic example of wedging, with clams steadily improving their position through time, eventually coming to dominate.

Gould and Calloway's detailed study of the fossil record revealed a different story. At times when clams thrived, so did brachiopods; they underwent bad times in unison, too. The long-assumed gradual replacement through competitive wedging did not take place; instead, the switch in dominance was "simply a result of different reactions to the greatest of all mass dyings— the Permian extinction." While the clams were virtually unaffected by the crisis, the brachiopods suffered badly. "Thus, clams got 'ahead' of brachs in this one geological moment and never relinquished their new incumbency."[4] Whether it was genes or luck that favored the clams in their unhampered march through the Permian debacle is, however, not immediately apparent from this statement.

A clear example of bad luck is seen in the history of the heath hen, a favorite prey of hunters in colonial America. Originally, its range stretched from Maine to Virginia, but by 1908 intensive hunting, combined with a loss of habitat through human population expansion, reduced the birds to only fifty individuals, isolated on the island of Martha's Vineyard, off the coast of Massachusetts. It was certainly bad luck for the bird to be so desired as a prize on the colonial table, but that's not the point here. A refuge of some sixteen-hundred acres was established to protect the remaining birds and to boost their population. By 1915, the project was gaining momentum, with two thousand birds in the refuge.

Then disaster struck, or rather a series of disasters. Fire, a hard winter, the inimical effects of inbreeding, and a poultry disease reduced the population to a mere eleven males and two females in 1927. The last bird was seen on 11 March 1932. The population succumbed not through bad genes, but through bad luck. In its original range, of many thousands of square miles, the species was virtually immune to extinction; reduced to a tiny, single population, it became vulnerable to the vagaries of the environment.

These and similar examples did not consign competition to oblivion in the realms of evolutionary biology, but they did encourage biologists to be aware of other factors in the history of life. Competition *is* important in evolution, as a component of natural selection; of that there is no doubt. Some of the best examples are the interactions between predator and prey, such as an increase in power of crab claws, matched step by step with an increase in the thickness of the shells of their prey, including molluscs; in the chemical warfare between insects and the leaves they eat; and in the greater grinding power of grazers' teeth and the defenses developed by grasses, such as the inclusion of crystals of silica, or phytoliths. There is a struggle for existence, in which species may be locked in battle just as I described, or compete with a similar species for access to the same resources, whether they be nutrients in the ground or meat on the hoof. But the wedge metaphor probably overplays the extent to which every potential niche is occupied.

Our question is: How is this to be encompassed in the pattern of extinction we know to exist? That pattern includes two phases, the intervals of background extinction, in which species disappear at a low rate, and episodes of high rates of extinction, including major biotic crises. Most biologists agree that the prevailing force in times of background extinction is natural selection, in which competition plays an important part. What of the bursts of higher rates of extinction? Is there simply a quantitative difference between periods of background extinction? Do marine regressions, climate cooling, and the effects of asteroid or comet impact tighten the screws of competition as times get tough? In other words, is mass extinction merely background extinction writ large? Is most extinction, including mass extinction, principally the result of bad genes? Until recently, the answer to these questions would have been an unequivocal yes.

One winter weekend two decades ago at the Woods Hole Marine Station in Massachusetts, David Raup, Stephen Jay Gould, and a few of their colleagues were mulling over the issue of extinction. Raup, a geologist who prefers to do his field work in front of a computer screen, proposed a statistical experiment to see if extinction patterns might be merely the result of stochastic processes, or chance. The human eye is rather poor at detecting the effect of chance, and much prefers to pull significance out of patterns, even where none exists. In a series of about a dozen research papers, Raup, Gould, joined variously by Jack Sepkoski, Thomas J. Schopf, and Daniel Simberloff, tried to see whether an artificial world of ecological communities (effectively, a computer simulation program) would exhibit patterns of background and mass extinction when governed only by the random chance of individual species' survival.

Although they finally concluded that randomness alone was insufficient to account for the shape of the history of life in the Phanerozoic, and that some selective forces must be operating, they did see patterns similar in form if not in magnitude to what is found in the fossil record. Moderate extinction episodes did occur, by the chance, simultaneous extinction of many species. But huge tracts of time would be necessary for extinction episodes of the magnitude of the end-Cretaceous or the end-Permian to occur by chance. Bad luck, in the sense of making the wrong call in a heads-tails choice, therefore cannot be the sole cause of a species' demise in a mass extinction event. But Raup and his colleagues' exercise was part of the move toward the realization that "bad genes" were not the sole explanation of the pattern of life. There was some combination of selection and bad luck.

As part of the research effort inspired by the Alvarez suggestion in 1979 of a link between asteroid impact and mass extinction, the University of Chicago paleontologist David Jablonski began to uncover the nature of that selection. More specifically, he compared the pattern in background and mass extinction periods. During background extinction, several factors conspire to protect a species from extinction. As is done with much of this research, Jablonski worked with fossil marine species, but the same principles hold for terrestrial species, too.

Species that are geographically widespread resist extinction in times of noncrisis. We saw that the small, isolated population of heath hens after being decimated by human hunting was vulnerable to chance events. Had the birds maintained their previous large range, the fires, hard winters, and bouts of pestilence would have killed local populations, but not the species. This makes sense. Second, marine species that send their larvae far and wide (drifting with the currents) also resist extinction for similar reasons. A group of related species, what biologists call a clade, resists extinction if it contains many species rather than few. Here, the chance disappearance of a few species is likely to be more dangerous to the survival of a clade that has only three species, say, than one that has twenty.

When Jablonski examined the fate of mollusc species and species' clades across the end-Cretaceous extinction, he saw a very different picture. None of the above rules applied. The only rule he could discern applied to groups of related species, clades. Again, geographic distribution played a part. If a group of species occurred over a wide geographic range, they fared better in a biotic crisis than those which were geographically restricted, no matter how many species made up the clade. "During mass extinctions, quality of adaptation or fitness values . . . are far less important than membership in the particular communities, provinces, or distributional categories that suffer minimal disturbance during mass extinction events," wrote Jablonski.[5] This was a landmark result, because it was the first that clearly indicated that the rules changed between background and mass extinction. Biotic crises are *not* simply background extinctions writ large.

This makes sense, because in the history of life, many successful species or groups of species have met abrupt ends in mass extinctions. The dinosaurs and ammonoids had been dominant in their realms for more than a hundred million years, and were as diverse as they had ever been when they were snuffed out at the end-Cretaceous extinction. There is no indication that the mammals were better adapted in any way than dinosaurs, which they subsequently replaced as the major terrestrial tetrapod group. And in the oceans, reef communities have been transformed periodically, as existing dominant organisms were wiped out, coinciding in four cases with the major extinction crises. After each devastation, reefs reappeared, sometimes with calcareous algae dominating, sometimes bryozoans, sometimes molluscs, and sometimes corals. The coral reefs with which we are familiar are simply the temporary inhabitants of that adaptive zone. None of these is superior to the others in any obvious way.

Just as we saw with the loss of large numbers of phyla subsequent to the Cambrian explosion, there are no examples of major faunal changes through mass extinctions in which the winners can be assessed as having been adaptively superior. We are judging from afar, of course, and it is possible that adaptive superiority did play a part, but if it did, the elements are too subtle for us to see. Nevertheless, as Raup points out, "The sad truth is that there is no hard evidence, other than the fact of the extinctions, for the inferiority of the victims."[6] Moreover, there is a strong argument for suggesting that adaptive superiority—in the day-to-day Darwinian sense—simply cannot have been a factor in major biotic crises. Natural selection operates cogently at the level of the individual in relation to local conditions—the impact of competitors and prevailing physical conditions. It is a powerful response to immediate biological experience. But it cannot anticipate future events. And it certainly cannot anticipate rare events.

If the average longevity of an animal species is four million years (and this is probably an overestimate, based as it is on those species which are successful enough to leave a fossil record), and if extinction bursts occur on average no more than every twenty-six million years, then most species never experience such bursts. The mass extinction episodes—the Big Five—are rarer still, making them invisible to natural selection. As Raup puts it, "[The] likely causes of extinction of successful species are to be found among stresses that are *not* experienced on time scales short enough for natural selection to act."[7]

The consequences of this are profound. As Gould has stated, "If mass extinctions . . . [are caused by agents] . . . so utterly beyond the power of organisms to anticipate, then life's history either has an irreducible randomness or operates by new and undiscovered rules for perturbations, not (as we always thought) by laws that regulate predictable competition during normal times."[8] The consequences of mass extinction are therefore going to look random to some extent, especially if the measure used is the quality of adaptation.

I stated in the opening paragraph of this chapter that the effect of mass extinctions was to reformulate the cast of characters on life's stage, and asked whether this was determined by bad genes or bad luck. By now it is obvious that bad genes is not the answer. Let's look again at the end-Cretaceous extinction. Acknowledging that mass dying occurred in every biological realm, I'll concentrate on the terrestrial tetrapods, because they illustrate well the point to which we are moving. William Clemens at Berkeley has drawn up lists of these beasts, and their fate. There were 177 genera of fossil mammals, amphibians, reptiles, and fish. Fifty genera vanished in the biotic crisis, twenty-two of them dinosaurs.

For dinosaurs, this spelled the end; not a single genus of these remarkable beasts survived. By contrast, only one genus of placental mammal disappeared, while marsupial mammals (often considered more primitive) suffered the loss of three quarters of their genera. On the face of it, placental mammals seem to have been doing something right, while dinosaurs and marsupial mammals were doing something wrong. For 140 million years, the terrestrial sphere was dominated by a dinosaur way of life (there were about fifty genera alive at any one time), which included laying eggs as a means of reproduction. After the end-Cretaceous extinction, placental mammals came to dominate. The internal development of young now was the most common means of reproduction in the tetrapod realm. The method used by marsupial mammals, a short period of internal development, followed by a semi-internal one in a pouch, is uncommon in numerical terms.

Under the Darwinian paradigm of progressive improvement, the obvious inference is that placental reproduction is superior to the amniote egg or the marsupial pouch. But if we ask why the placental mammals survived through the extinction while the dinosaurs didn't and the marsupial mammals suffered poorly, a different answer emerges. It's unlikely that dinosaurs fell victim for reasons of restricted geographical distribution; they were widespread. But most of them were large creatures, and there is some correlation between large body size and vulnerability to extinction. The reason may have to do with demography, such as small populations, large home range, slow reproductive rate, and so on. There were probably other factors involved, too, almost certainly having nothing to do with individual species adaptations. What of the marsupial mammals? According to Clemens, their fate was sealed through geographic isolation, in Australia and South America. Geographic distribution, not adaptive inferiority, was at work here.

For more than a hundred million years, mammals had coexisted with dinosaurs, mainly as small, nocturnal, tree-dwelling creatures. Their ascendancy after the end-Cretaceous extinction was due, surely, as much to good luck as to good genes, possibly entirely so. This scenario, played out by different groups of organisms, in all biological realms, and at many points in the last six hundred million years of Earth history, gives an entirely different cast to how we perceive life's flow. Mass extinctions were once considered to be mere interruptions of that flow. Yes, the nature of the biota often shifted following these events, but that was considered a result of a tightening of the screw of competition. Now it is clear that different rules apply during these times, casting Darwinian competition aside temporarily, and playing on forces for which species are unprepared, indeed, cannot be prepared. Mass extinctions do not merely reset the clock of evolution, jolting it back for a while; they change the face of the clock. They create the pattern of life.

As David Jablonski put it, "It is the alternation of background and extinction regimes that shapes the large-scale evolutionary patterns in the history of life."[9] During times of background extinction, Darwinian natural selection operates, creating evolutionary novelties, shaping adaptation, creating the fit between organisms and their environments. Genes and luck both play roles in this regime, with bad genes perhaps predominating over bad luck in terms of extinction. During mass extinction, Darwinian rules are suspended, and species survive or succumb for reasons unrelated to their adaptations. Here, bad luck is dominant in consigning species to evolutionary oblivion.

Mass extinctions, then, restructure the biosphere, with an unpredictable set of survivors finding themselves in a world of greatly reduced biological diversity. With at least 15 percent, and as much as 95 percent, of species wiped out, ecological niches are opened, or at least made much less crowded than they once were. It is a time of evolutionary opportunity offered to a lucky few.

I described earlier the pattern of diversity through the Phanerozoic as an upward trend from the Cambrian explosion to the present, interrupted by major collapses. This is something of a simplification, as we shall see. It is also relatively new, at least in the pages of the paleontological literature. During the 1970s, the pattern of the Phanerozoic was the subject of debate and strong disagreement. For instance, James Valentine at Berkeley, argued for a rapid and ever accelerating increase in animal species diversity, based on his reading of the fossil record. David Raup, by contrast, suggested it was a pattern of rising to an early maximum, followed by decline. He based his conclusion on a statistical analysis of the record. A compromise was proposed by Richard Bambach, of the Virginia Polytechnic Institute, again based on a reading of the fossil record.

It may be thought curious that the same record can be interpreted in so many different ways, but, as I keep emphasizing, the imperfection of the record makes a simple reading of it impossible. Many assumptions have to be made, and allowances for the deterioration of the record the farther back you look. The differences of opinion were resolved when, joined by Jack Sepkoski, the three protagonists got together to sort out the problems of their different approaches. In October 1981 the four researchers jointly published what is known in the trade as the "kiss-and-make-up paper." In it, they described a general, but stepwise, increased diversity through the Phanerozoic, something like the middle ground that Bambach had proposed. As usual, this was based principally on analysis of the marine fossil record, but something similar is seen among terrestrial vertebrates and plants, in the pattern, if not in the timing.

As with the mass extinctions themselves, the duration of the biotic rebound from these events differs considerably, from a few thousands of years after the end-Cretaceous to many millions following the Permian extinction. Many factors must be involved, one of which is the nature of the animals that survive the crisis, an issue that Sepkoski has investigated in detail. He identifies three phases of diversification through the Phanerozoic. First is the Vendian-Cambrian phase, which includes the events of the latest Precambrian, the Cambrian explosion itself, and the stabilization through the late Cambrian. Second is the late Paleozoic phase, beginning with the Ordovician radiations, which generated a tremendous and rapid rise in diversity, more than tripling the count of animal genera and families in the marine realm. This was followed by some 200 million years of apparent equilibrium, punctuated here and there by diversity collapses of various magnitudes. The last phase, the Mesozoic-Cenozoic, includes the rebound from the massive end-Permian event and the continued expansion through to the recent.

This stair-step pattern is not the result of concerted diversification, or stasis, within the entire community of species, but rather within discrete groups, or evolutionary fauna, as Sepkoski calls them. The Vendian-Cambrian phase of diversification was driven by the so-called Cambrian fauna, including trilobites, molluscs, and echinoderms. These species expanded rapidly early, but then tailed off as the next phase began, and subsequently played only a minor role. The trilobites, for instance, disappeared altogether at the end-Permian extinction. The late Paleozoic phase was carried by brachiopods, corals, bryozoans, and cephalopods, the Paleozoic fauna. Although they were responsible for the tremendous rise in diversity in this phase, they had been on the scene much earlier, having arisen mostly in the Cambrian. Extinction events at the end of the Cambrian gave them their chance. The Mesozoic-Cenozoic phase was the product of what Sepkoski calls the Modern fauna, and included bivalve and gastropod molluscs, crustaceans, echinoids, and true fish. Again, many of these groups evolved in the early Paleozoic, where they played an inconspicuous part. Having survived the end-Permian extinction, their number and diversity exploded, eventually leading to the life we see in the modern oceans.

The dynamics of diversity rebound are complex and can involve an increase in numbers of species within ecological communities, the origin of more and different kinds of communities, and the development of specialist communities; that is, those which occur only in restricted habitats. A common element in the early phase of recovery, however, is a spike in the turnover of species; that is, the life span of species is temporarily shorter than usual. This implies some kind of evolutionary instability, the result, perhaps, of an unleashing of evolutionary experimentation in relatively open ecological space. In addition, there are far more evolutionary novelties generated in the early phase of recovery, probably for the same reason. As we saw, the extent of novelty produced never equaled that of the remarkable Cambrian explosion.

An evolutionary rebound following a devastating collapse in diversity is not surprising: ecosystems are suddenly more open, inviting invasion by new species. But the upward trend in diversity, particularly in the last phase, the Cenozoic, has to be explained. Sepkoski's view is that it results from the makeup of the fauna and their basic ecological strategies. Valentine has a different explanation, which relates to the configuration of the continents.

Species of the Cambrian fauna were ecological generalists in many ways, which is consistent with the low diversity in this phase. Compared with specialists, which have narrow ecological requirements, generalists are tolerant of a relatively wide range of environments. Specialist species are therefore plentiful, as each occupies a small segment of habitat; generalist species are far fewer, because each can spread over a much wider territory. The boost in species diversity in the second phase, the late Paleozoic, may be the result of the appearance of many more specialist species. The 200-million-year plateau of diversity may imply that the global ecosystem had reached some kind of equilibrium, suggests Sepkoski; in other words, the ecospace was full. With the Modern fauna, predation became a much more important aspect of community dynamics. Ecologists know from their mathematical models, and from some empirical observations, that predation can increase diversity, often because predators prevent any single species or few species from dominating the system. This, on a global scale, may underlie the great rise of diversity in the Cenozoic. "Thus," suggests Sepkoski, "levels of biodiversity in the marine realm would seem more a function of the identity of the players than the act of the drama."[10]

Valentine's explanation points to the habitats, rather than habits, of the species. During the past 200 million years, the supercontinent, Pangea, has been breaking up under the influence of tectonic movements. During the past hundred million years, the continents as we now know them began to take shape; there has been an increase in the number of individual landmasses, ultimately arrayed from pole to pole. Such a configuration maximizes the number of available habitats, in both the marine and terrestrial realms.

Whatever mechanism is operating (perhaps both), *Homo sapiens* evolved amid a high point of global biodiversity. Waves of extinction of large land mammals during the past two million years—perhaps driven by glacial periods—has reduced that diversity to a degree, but not drastically so. We are but one of millions of species here on Earth, product of half a billion years of life's flow, lucky survivors of at least twenty biotic crises, including the catastrophic Big Five.

The Engine of Evolution

FOR MORE than half a billion years, evolutionary processes have constantly generated variations on themes established in the Cambrian explosion. It was a time of unmatched evolutionary innovation.

Homo sapiens came into being at a point in earth history that boasted virtually the richest diversity of life forms that has ever existed. We may be the highest expression of life's arrow of evolution.

Sentient species that we are, *Homo sapiens* is able to understand the shape, extent, and value of Earth's biodiversity. Indeed, we have a responsibility, as well as a self-interest, to value it.

6

Homo sapiens, the Pinnacle of Evolution?

T HE ANSWER to the above question appears self-evident. Yes, of course we are. In the penultimate chapter of *The Origin of Species,* Darwin wrote, "As natural selection works solely by and for the good of each being, all corporeal and mental endowments will tend to progress towards perfection."[1] *Homo sapiens,* since its origin some 150,000 years ago, has come to occupy every continent, with the exception of the hostile wastes of Antarctica, and even there we have a toehold. This surely attests to our corporeal endowment, as we have adapted to these many different environments. And there is no question about our mental endowment, which is unmatched in all of nature. We are intellectually analytical, we are artistically creative, and we have invented ethical rules by which society operates. No one can doubt that our species has advanced toward, if not perfection, then a high point—the highest point—among the diversity of life on Earth. We *are* the pinnacle of evolution. Or are we?

Anthropologists and biologists have struggled with this issue for a very long time, and the resolution has never been simple.

We feel ourselves unique in the world of nature, and of course we are: each species is unique, by definition, so that doesn't help much. We are but one species among many millions in today's world. However, we feel ourselves special, among this exuberant diversity of life, because we have an unmatched capacity for spoken language and introspective consciousness, and we can shape our world as no other species can. We judge this to place us on the top of the heap. Before the fact of evolution was demonstrated, beginning with Darwin in the mid-nineteenth century, we considered *Homo sapiens* to have been placed on the top by Divine Creation. In the Darwinian world, our species was said to have achieved its ascendancy through the natural selection of our special qualities. The intellectual context changed, but the outcome was the same. We judged ourselves to be the pinnacle of the world of nature.

This assessment brings two assumptions with it—one implicit, the other explicit. The implicit assumption is that the evolution of *Homo sapiens* was an inevitable outcome of the flow of life, in the unfolding of evolution. The explicit assumption is that the qualities we value in ourselves as a species are indeed superior in some way to the rest of the world of nature. Through evolutionary time, life became ever more complex, producing an arrow of progress. As Darwin stated in the above passage, by means of natural selection life "will tend to progress towards perfection." We are the tip of the arrow of progress, the expression of perfection.

Here, I will explore three aspects of this issue. First, I will discuss scholars' efforts over the millennia, from Aristotle to modern anthropologists, to place *Homo sapiens* appropriately in the world of nature. Second is the question of the inevitability of our species in the flow of life. If the history of life were set back to its beginning and allowed a rerun, would *Homo sapiens* emerge a second time among the great diversity of life? Last, I will ask whether the world of nature really has become more complex through evolutionary time. Is there an arrow of progress in the workings of natural selection?

Man's view of Man in the world of nature has changed over the centuries, reflecting the constant shift in scholarly context. Only in the relatively recent past have anthropologists begun to discuss human origins as they would the origin of oysters, cats, and apes. Nevertheless, a desire to maintain a boundary between us and our biological relatives can, even now, be discerned in some scholars' theories on human prehistory, particularly in the matter of the origin of modern humans, people like you and me.

In the eighteenth and early nineteenth centuries, scholars saw order in nature in the form of the Great Chain of Being, which had roots in Aristotle's worldview. It was a "sacred phrase," observed Arthur Lovejoy, a Harvard historian of science who made the classic study of the concept, published in 1936. And it played a part "somewhat analogous to that of the blessed word 'evolution' in the late nineteenth century." From the simplest forms of life, the bacteria, to the most complex, *Homo sapiens,* nature was arranged in regularly graded intervals, which reflected the orderliness of creation. The chain was meant as a description of the world of nature as it has been since creation and as it always would be.

If the chain of being had fully reflected Aristotelian perfection, and later scholars' expectations of order in nature, then it would have been unbroken; there would have been no gaps in the gradation of the natural world. However, there were large apparent gaps; namely, between minerals and plants, between plants and animals, and, most embarrassing of all, between apes and humans. So influential was the theory that when, between 1736 and 1758, Carolus Linnaeus established the basis of zoological classification—his *Systema naturae*—he postulated the existence of a primitive form of human, *Homo troglodytes,* that filled the chasm between humans and apes. *Homo troglodytes* was said to live in forests, to be active only at night, and to communicate only in hisses. Contemporary explorers often returned from Africa with fantastic tales of half-ape, half-human creatures, behaving in this way; they had "seen" what they expected to see, according to prevailing theory.

With the advent of Darwinian theory, the source of order in the world was viewed very differently. Rather than the product of creation, order was the outcome of history, or "descent with modification," as Darwin put it. All organisms shared common roots, connected variously through the unfolding of evolution. And this, of course, included humans. Nevertheless, the scholars' view of the world remained remarkably similar to the pre-evolutionary position. That is, *Homo sapiens* was assumed to represent the ultimate product of evolution and to be separate from the rest of nature in some important sense, with a gradation of increasing superiority through the geographical races, from Australian to European.

For instance, Alfred Russel Wallace, the co-inventor of the theory of natural selection, believed that evolution had been working "for untold millions of years . . . slowly developing forms of life and beauty to culminate in man."[2] The Scottish paleontologist Robert Broom, who in the 1940s and 1950s helped pioneer the search for early human fossils in South Africa, essentially agreed with Wallace, stating the following in 1933: "Much of evolution looks as if it had been planned to result in man, and in other animals and plants to make the world a suitable place for him to dwell in."[3] Broom clearly saw humans as special and separate, and the rest of the natural world ours to exploit as we please. Broom's was not an isolated opinion; it accurately portrayed contemporary thinking. Anthropologists of the time were in awe of the size and power of the human brain, and saw it as the biological endowment that placed us on the top of the heap, lord of all. Human progress through prehistory, according to the prominent British anthropologist Sir Arthur Keith, had been "a glorious exodus leading to the domination of earth, sea and sky."[4]

Examples of what by today's standards we would condemn as blatant racism were legion in scholarly writings of the early decades of the century, which placed in an evolutionary framework what had been seen as the product of creation in earlier times. One citation will suffice by way of illustration. In his *Essays on the Evolution of Man,* the eminent British anatomist Sir Grafton Elliot Smith wrote the following in 1923:

> The most primitive race now living is undoubtedly the Australian, which represents the survival with comparatively slight modification of perhaps the primitive type of the species. Next in order comes the Negro Race, which is much later and in many respects more highly specialized, but sharing with it the black pigmentation of the skin, which is really an early primitive characteristic of the Human Family that primitive Man shares with the Gorilla and Chimpanzee. After the Negro separated from the main stem of the Family, the amount of pigmentation underwent a sudden and very marked reduction;

and the next group that became segregated and underwent its own distinctive specializations was the Mongol Race. After the separation of the Mongol Race there was a further reduction of pigment of the skin; and from this white division of mankind the Alpine Race first split off the main stem, which ultimately became separated into the Mediterranean and Nordic Races, in the latter of which the reduction of pigment was carried a stage further to produce the blondest of all human beings.[5]

Overt racism of this kind disappeared from texts by mid-century, with a curious effect. Viewed as more primitive than white Caucasians, the "inferior races" formed something of a bridge between the ultimate expression of *Homo sapiens* and the rest of the animal world. When all races were regarded as being equal, that bridge disappeared, and a gap opened up, making modern humans appear even more separate from the world of nature. Anthropological writing focused on the traits we admire—big brains, language, and technology—and the evolution of these was assumed to have been part of our history from the very beginning of the human family. We were propelled by the force of evolution, special from the start, to a special place in the world.

So special, in fact, that Julian Huxley, grandson of Thomas Henry Huxley, suggested that humans should be classified entirely separately from the rest of the world of nature. "In a truly evolutionary view, [humans] constitute a radically new and highly successful dominant group, evolving by the new method of cultural transmission," he wrote in 1958. "Man . . . must be assigned to a distinct grade, which may be called the Psychozoan."[6] Nonhuman animals, plants, and all other organisms are classified in various kingdoms of their own, each in company with many species. In Huxley's view, humans would be the sole occupants of their own biological kingdom.

I would not argue that the cultural realm that humans create is not special in the world, but since Huxley made his suggestion we have come to appreciate much more clearly the abilities of other creatures, notably our closest relatives, the great apes. Our language and cultural capacity was said to separate us from the world of nature in many ways. Only humans use tools, it was said; only humans have a sense of self-awareness; only humans can elaborate culture; only humans have symbolic language. Researchers doing naturalistic studies, such as Jane Goodall and

Dian Fossey, have eroded the human-animal boundary that we so assiduously constructed from these supposedly unique attributes. Apes do use tools; they do have a culture of sorts; they are self-aware; and, although it is a controversial realm of research, there's a good probability that apes, unable to produce spoken language, have the capacity to understand and manipulate the symbolism that spoken language embodies. We are not so special, after all.

Moreover, the long-held notion that our ancestors were essentially human from the beginning of the human family has also crumbled. From the evidence of the fossil and archeological records, and from data from molecular biology, we now know that, although the first human species evolved about five million years ago, the cherished attributes of an enlarged brain and technological ability did not appear until about 2.5 million years ago. For a long time in our prehistory, we were bipedal apes, and no more. The human family underwent the kind of adaptive radiation we see in many groups of large, terrestrial vertebrates: from a founding species of the group, many new species evolved, a bush with slight variants on the original theme. The advent of brain expansion that occurred in the genus *Homo,* the lineage of which we are the sole surviving member, produced an important change. We were no longer merely bipedal apes; we were bipedal apes that had begun a new adaptation, that of a primitive form of hunting and gathering. For almost 2.5 million years, the brain gradually got larger, and the hunting-and-gathering adaptation became ever more developed. I've no doubt the capacity for spoken language built gradually, too.

From this perspective, *Homo sapiens* may be blessed with special traits, but we are not separated from the rest of the world of nature by an enigmatic gulf; we are joined to it by a succession of ancestors in whom these traits developed gradually. I find it interesting—amusing, even—when I hear some of my anthropology colleagues speak of human prehistory in a very different way. Modern humans—with our cherished behavioral traits—arose very recently and abruptly, they argue. All human species prior to modern *Homo sapiens* were much more like apes than like humans; they were like chimpanzees in their cognitive capacity, their communicative capacity, and their subsistence activity. How this can be said of a species that had a body form like ours and a brain capacity very close to, and sometimes exceeding, our own, escapes me. The only conclusion I can reach is that, like the anthropologists in the early part of the century who could not bring themselves to contemplate a human evolutionary origin with anything as base as an ape, some of my contemporaries fervently wish to maintain the boundary between us and the world of nature. Their wish is futile.

We are therefore, I believe, neither as special as has been claimed nor as separate from our biological relatives as many appear to wish. But we still could be the pinnacle of evolution: the inevitable and most complex product of this creative process.

"Life, if fully understood, is not a freak in the universe—nor man a freak in life," the French Jesuit priest, philosopher, and paleontologist Pierre Teilhard de Chardin wrote in his most famous book, *The Phenomenon of Man,* in 1959. "On the contrary, life physically culminates in man, just as energy culminates in life." Teilhard de Chardin's certainty was based on his philosophy of a spiritually guided flow of energy in the universe, the foreordained endpoint inevitably being Man. We have to address the same question, but our point of reference is what we may infer from an understanding of the physical flow of life, as we saw it earlier in this book. The matter of the inevitability of our evolution is somewhat intertwined with the question of a rise in biological complexity through the history of life, but I will pull the two issues apart and discuss them separately, bringing them together again at the end.

Charles Lyell's dictum "The present is the key to the past" may be a little optimistic, because we are constrained by our experience of the present. For instance, I have visited Lake Turkana, in northern Kenya, more times than I can now count, looking for ancient human fossils in the sandstone sediments that compose its eastern and western shores. It is an enormous lake, shaped like a dog's leg, some 150 miles north to south and perhaps 30 miles east to west. No one who sees it fails to be impressed, vibrant as it is in its stark setting of sun-baked, arid terrain. Since my first visit, in 1968, the water level has dropped some sixty feet, making the walk from our camp to the shore for morning ablutions a great deal longer. This drop didn't surprise me, because I had read accounts by nineteenth-century travelers, which record substantial fluctuations in lake level. Nevertheless, it remains a giant lake by any standards.

Fluctuations of this magnitude in the lake's recent history are, however, minor compared with changes it has undergone in prehistory. Frank Brown of the University of Utah has surveyed the lake and its surroundings in almost two decades of work there, charting its existence over the past four million years. His findings are breathtaking. At times the present lake was dwarfed by a much bigger version. Just ten thousand years ago, for instance, the waters covered ten times the area they do now. At other times no lake existed in the Turkana Basin; there was merely a river running through it. Although I know what he says to be true, I find it all but impossible to encompass in my mind, so powerful is the lake's presence before my eyes today. I simply cannot imagine its not being there.

The same problem confronts us when we try to contemplate a world without *Homo sapiens*. Our self-awareness impresses itself on us so cogently, as individuals and as a species, that we cannot imagine ourselves out of existence, even though for hundreds of millions of years humans played no part in the flow of life on the planet. When Teilhard de Chardin wrote, "The phenomenon of Man was essentially foreordained from the beginning," he was speaking from the depth of individual experience, which we all share, as much as from religious philosophy. Our inability to imagine a world without *Homo sapiens* has a profound impact on our view of ourselves; it becomes seductively easy to imagine that our evolution was inevitable. And inevitability gives meaning to life, because there is a deep security in believing that the way things are is the way they were meant to be.

How does this yearning for inevitability measure up to what we know of the history of life?

We must answer this on two levels, global and local. The global level involves the major shifts in the history of life, especially the biotic crises that punctuate that history, and in particular the Big Five. On a local level are environmental changes that have influenced human prehistory.

When we contemplate the history of vertebrate life in broad view, we see an evolution from fish to amphibians to reptiles to mammals to primates to humans, an apparent progression from the primitive to the advanced. "If evolution has moved along such a linear path of progress, how could the story have un-folded in any basically different way?" Stephen Jay Gould asks rhetorically. The progression appears to have a certain logic to it, so much so that, says Gould, we have a tendency to assume that "whatever is, is right."[7] But we know very well that this is proba-bly untrue.

We now understand that the Cambrian explosion was a period of unprecedented evolutionary experimentation, yielding a vast range of body plans, or phyla, which became the basis for all of the rest of the history of life. Perhaps that range of body plans, as many as a hundred in all, exhausted the range of innovation available to evolution, so that if, by some magic, the whole pro-cess were run through again, the same phyla would be present. But there is no theoretical or empirical argument to back this up. True, there are some constraints in possible biological architec-ture, usually relating to biomechanics (there are no organisms with wheels, for instance). But many of the Cambrian animals already strain our experience of life, and our credulity, too. So, although there is not an infinity of forms, the Cambrian bestiary almost certainly was just one of many such possible worlds. Re-wind the tape and run it again—to apply Gould's metaphor for this thought experiment—and other worlds might appear.

So what? So one of the body plans that arose in the Cambrian explosion was the phylum chordata, the root source of verte-brates. A second such explosive production of life forms might fail to produce chordates. And, as we know, if a phylum does not appear in the first creative explosion, it would never do so. A world flowing from a rerun of life's tape might be a world with-out backbones: no fish, no frogs, no lizards, no lions—no us.

But the Cambrian explosion *did* yield a chordate, a modest organism that Charles Walcott named *Pikaia gracilens*. Judging by its rarity among the Burgess Shale fauna, it was not an important player in Cambrian life. Exist, it did, however, so perhaps I may be accused of playing mind games of no intellectual value. True, we can't rerun the Cambrian explosion, so we will never know the truth of this possibility. We can, however, look at the gauntlet of the history of life through which chordates survived, making our existence possible.

One of the principal themes to emerge from the previous sec-tion was that, although Darwinian selection has been important in shaping organisms' adaptations—fitting them to the exigencies of life—biotic crises temporarily suspended Darwinian rules, replacing them with another set, about which we are only just beginning to learn. What we do know, however, is that chance played a disconcertingly powerful role in determining which species survived and which did not. And this applied to the fate of the hundred or so basic body plans that were produced during the Cambrian explosion: the hand of randomness was present in whittling them down to the thirty that formed the basis of modern life. *Pikaia* did survive that first of life's grand lotteries, forming the root of all vertebrates, including *Homo sapiens,* but it might not have. Survival was a matter of historical contingency, not good design, a theme that Gould has developed persuasively in recent years, and presented in detail in *Wonderful Life.*

From humble beginnings, vertebrates evolved, first as fish, then as terrestrial tetrapods, a transition that flowed from a minor group of fish, the lungfish-coelacanth-rhipidistian group. "Replay the tape, expunge the rhipidistians by extinction, and our lands become the unchallenged domain of insects and flowers," notes Gould.[8] But the transition did take place, and eventually mammals evolved, more or less at the same time as the dinosaurs, at the end of the Triassic period, some two hundred million years ago. For more than a hundred million years, mammals played a minor ecological role while the Terrible Lizards dominated terrestrial life. And dinosaurs thrived until their sudden demise at the end of the Cretaceous, very probably victims (in company with millions of species in all ecological communities) of the aftermath of a comet's collision with Earth. Had the comet missed, dinosaurs might still be the dominant terrestrial tetrapods, with mammals still scurrying around as rat-sized creatures in the crepuscular worlds of dense forests.

But the comet did make impact; the dinosaurs did vanish (leaving birds as their sole descendants); placental mammals survived with relatively modest losses, and included a small creature, *Purgatorius,* the earliest primate. But what if *Purgatorius* had not survived . . .

I can be accused of playing the "if . . . what then?" game, to which there is no real answer. But I believe that Gould has been correct to raise our consciousness to the role of contingency in life's flow, although I suspect he pushes the argument too far.

Viewed dispassionately, however, mass extinctions must be seen to have exerted a major role in shaping the history of life, and in unpredictable ways, too. That unpredictability means that the progress from fish to us was simply what happened through evolutionary history, not what *must* have been. Read like this, the history of life implies that there was no inevitability in the evolution of *Homo sapiens.* We are, as Gould says, "a wildly improbable evolutionary event."[9] That may be difficult for us to accept, but it is surely true.

History at the local level leads to much the same conclusion we see on the global scale. Twenty million years ago, lush forest formed a belt across the African continent, from west to east. A dozen species of ape thrived then, perhaps more. Monkeys were not yet an important part of the ecological community. Over the next fifteen million years, dramatic changes occurred across the continent, driven primarily by geological forces deep under ground. By then—five million years ago—only a handful of ape species remained, monkeys were numerous, and the human family had evolved somewhere in East Africa. Two-and-a-half million years later, global temperatures plunged, and massive ice caps formed at both poles. The genus *Homo,* the large-brained lineage that led to us, evolved. That's the history, in broad outline. Now let's see what happened, and why.

The geological forces that changed the face of the continent flow from the separation of two tectonic plates, running from the Red Sea in the north through Mozambique in the south. Upwelling of magma forced two giant "blisters" to rise, the Ethiopian Dome and the Kenyan Dome, each at least six thousand feet in altitude. As the plates continued to move apart, the overlying continental rock became ever more strained. Eventually, about ten to twelve million years ago, it gave way, with faulting and further uplift forming a long, deep valley, called the Great Rift Valley, one of the few geological features that can readily be seen from space.

The ecological effects of these physical changes were dramatic. The west-to-east prevailing winds that had sustained the continentwide forests with moisture picked up from the Atlantic now faced a barrier—the uplift associated with the Rift Valley. Forced higher, the winds dropped their moisture, throwing the land to the east of the valley into rain shadow. Deprived of moisture, the forests began to shrink and fragment, and were replaced by open woodland and, eventually, grassland. Where once there had been a more or less continuous habitat of forest, there were two ecological zones, persistence of forest in the west and the newly formed mosaic of woodland and grassland to the east. This mosaic was further enriched by the Rift Valley itself, which provided habitats from cool highlands to hot, dry desert on the valley floor. Biologists recognize that a richly patterned mosaic habitat is a powerful engine of evolution, because it presents many opportunities for different kinds of adaptation. The high biodiversity in the region today is testimony to that.

The ecological consequences of the formation of the Rift Valley appear to have been instrumental in the evolution of the human family. Humans share a common ancestor with modern chimpanzees, which diverged about five million years ago. To the west of the valley, apes could continue living the lives to which they were adapted; that is, in heavily wooded and forested terrain. This is where modern chimpanzees and gorillas live. The land to the east of the valley was no place for apes, with its forests rapidly disappearing as rainfall levels diminished. One very persuasive theory for the origin of bipedalism, the feature that established the human family, is that it was an adaptation for more efficient locomotion between widely distributed food sources. There are other theories, too, but this one makes good biological sense, given the habitat changes of the time.

We know very little about the early prehistory of the human family, because the fossil record is virtually nonexistent prior to about four million years ago. (The date of five million years ago for the origin of the family is derived from a comparison of the genes of humans and African apes and from recently discovered fossils.) By some three million years ago, the record is good enough for us to see that several human species—that is, bipedal apes—coexisted in East and South Africa. This is exactly the pattern we see in the evolution of a new adaptation, in this case a novel mode of locomotion. Over a period of several million years, a cluster of variations on the original theme evolved, with new species originating and others becoming extinct. The fundamental adaptation of the family was that of relatively lightly built bipedal apes with small brains; their diet included more tough

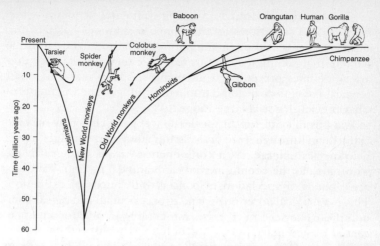

Family tree of primates. (Reprinted by permission of Blackwell Science/Roger Lewin)

foods than is normal for apes. These early members of the human family were species of *Australopithecus.*

Around 2.5 million years ago, another pulse of evolution appears to have occurred in the human family, yielding two different adaptations to a similar environment. In one, the original adaptation was further exaggerated, with the appearance of a bipedal ape that lived in drier environments and ate even tougher plant foods. The second comprised several major changes, including the development of a more athletic body, a bigger brain, and, significantly, the inclusion of meat in the diet, which was made possible by the first manufacture and use of stone tools. This was the genus *Homo,* of which there may have been several species early on. Both evolutionary directions can be seen as adaptations to a more arid environment.

Did something significant happen 2.5 million years ago to account for this burst of evolutionary activity? I think there was, and I point to Elisabeth Vrba's work on what she calls the turnover-pulse hypothesis. A South African anthropologist now at Yale University, Vrba spent many years analyzing the evolutionary history of antelopes in South and East Africa. She sees pulses of evolution at 5 million and 2.5 million years ago—periods, it transpires, of considerable global cooling. The effect of cooling was to change the environment, causing ecological communities to migrate, and sometimes dividing once continuous habitats into small fragments. Although animal species try to migrate with the habitat, maintaining conditions to which they are adapted, they are sometimes prevented from doing so. They may encounter physical barriers or become trapped in refugia, the fragments of what previously was a large range.

Two possible fates await species so trapped. They may become extinct; or, through genetic isolation, they may change, and a new species emerge. As a consequence, major changes in climate, such as the cooling periods at 5 and 2.5 million years ago, drive bursts of speciation in concert with bursts of extinction. This is Vrba's turnover-pulse, and it occurs in all ecological communities. The second of these two cooling events, for instance, caused evolutionary pulses not only in humans and antelopes, but also in rodents and many plant species. The effect is seen in Asia and Europe as well as in Africa. Notice that this implies that the engine of evolution is fueled by external environmental change, not by internal competition, which classic Darwinian theory suggests.

I said earlier that the human family originated about five million years ago, and that this estimate was based on genetic evidence and on indications from newly discovered fossils in Ethiopia and Kenya. The date is consistent with the turnover-pulse hypothesis, because of the marked global cooling that took place then. This provided a possible engine of evolution for the emergence of the human family, but we cannot be definite about it.

We should return to the "if . . . what then?" game, which is our current theme. If the tectonic forces that formed the Great Rift Valley had been absent, a carpet of forest might still lie across the continent, home to many species of ape but no bipedal apes. But let's allow the valley to develop, as it did, giving us small-brained, bipedal apes. If the cooling event at 2.5 million years ago had not taken place, the pulse of evolution would not have followed, with biological impetus for an adaptation to a more arid environment: no species of *Homo,* no us.

As long as the engine of evolution is largely fueled by external forces—capricious events in the environment—then nothing in evolutionary history is inevitable. Each species is a contingent fact of history. The events that led to the evolution of each species can be described, just as historical events can be described. But this does not imply that events *must* have unfolded as they did. They just did; that's all.

There are few issues more calculated to provoke strong disagreement among biologists than that of complexity. Or, more specifically, the idea that the process of evolution has resulted in greater biological complexity. On the face of it, the answer seems blindingly obvious: of course it has. Life started as simple, single cells, progressed through inconspicuous invertebrates, on upward through fish, amphibians, and mammals, and culminated in a species that is capable of contemplating its own evolution. Few would doubt that even the most primitive mammal is more complex than a single-celled organism. But if we restrict our question to multicellular organisms, what of the comparison between a mammal and a reptile? Mammals are more active, maintain a high body temperature, are usually more social, and, of course, have a much bigger brain. These attributes seem to bespeak greater complexity, and I have little difficulty accepting that. What, then, is the problem?

There are three: perceptual, philosophical, and practical. Because we are large terrestrial vertebrates, we tend to focus on the kind of biological world we inhabit and to value the characteristics we find there, such as sociality and intelligence. But trees are exquisitely attuned to their chemical environment, for instance, and moths can detect the presence of a potential mate some miles distant, guided by a few molecules of pheromone on the wind. Aren't these abilities to be valued, too?

That's the perceptual problem. The philosophical problem is complex in itself and relates to the frequent conflation of increase in complexity with an innate drive for progress in evolution. Biologists shrink from innate drives of this kind, and we've seen how the notion of progress in human evolution led in the past to a justification of referring to races other than our own as inferior or less developed. The practical problem is simply this: By what measure is the level of complexity to be determined? There is no obvious way to measure it.

I will apply some of these problems to the theme of this chapter: Are humans the pinnacle of evolution? For most of us the question refers to the human brain and its extraordinary abilities. (Our bodies, after all, have no special claim to being more complex than that of any other mammal's.) We can legitimately ask whether our brain represents the (present) culmination of evolutionary processes in its analytical and creative abilities and its self-awareness.

We should start with the theory of natural selection itself. Stephen Jay Gould, today's most prolific writer about evolution, is emphatic: "An implicit denial of innate progression is the most characteristic feature separating Darwin's theory of natural selection from other nineteenth-century evolutionary theories."[10] Gould suggests that one reason Darwin's theory was initially unpopular among his Victorian contemporaries was precisely because it denied progress in evolution: "It proposes no perfecting principles, no guarantee of general improvement; in short, no reason for general approbation in a political climate favoring progress in nature."[11]

It's true that natural selection works for the moment, responding to prevailing circumstances by shaping immediate adaptation. It is a local, not a global phenomenon, one in which there is no obvious arrow pointing toward greater complexity. Nevertheless, in *The Origin of Species* and other writings, Darwin was ambivalent about progress, sometimes denying it, sometimes embracing it. For instance, in a letter to the American biologist Alpheus Hyatt, Darwin wrote in 1872, "After long reflection I cannot avoid the conviction that no innate tendency to progressive evolution exists." And yet in the chapter on "Geological success of organic beings" in the *Origin* he stated that if species of a former era were put into competition with modern species, the ancient species "would certainly be beaten and exterminated." He then added, "I do not doubt that this process of improvement has affected in a marked and sensible manner the organization of the more recent and victorious forms of life, in comparison with the ancient and beaten forms; but I can see no way of testing this sort of progress."[12]

Although some of Darwin's contemporaries, such as Charles Lyell, balked at the idea of progress in nature, because it implied that humans, with their moral sense, were simply advanced apes, evolution and progress soon came to be synonymous. The British philosopher Herbert Spencer, inventor of the term "survival of the fittest," wrote in 1851 that "progress . . . is not an accident,

but a necessity." Somewhat later, Henry Fairfield Osborn, director of the American Museum of Natural History in the early
decades of this century, wrote, "Darwin's doctrine of evolution
. . . has been, and ever will be, the means of progressive evolution." He spoke for most of his colleagues.

Until about a decade ago, most biologists did not feel uncomfortable with speaking of an increase in complexity as an outcome of evolution, and using the term *progress* interchangeably
with *complexity.* Recently, however, a certain nervousness has
crept in, so that it is now acceptable to talk about complexity, but
not about progress. Progress, it is argued, implies some kind of
mysterious innate tendency for improvement, and that is considered too mystical. It was therefore a mark of courage when, in
1987, Matthew Nittecki, of the Field Museum in Chicago, organized a conference on evolution and progress. All but one
speaker, the geneticist Francisco Ayala, denied its reality. He
stated: "The ability to obtain and process information about the
environment, and to react accordingly, is an important adaptation because it allows that organism to seek out suitable environments and resources and to avoid unsuitable ones." This adaptation has become more refined through evolutionary history, he
said. I will return to this important point.

Gould was the most outspoken in denying progress, asserting
that it is "a noxious, culturally embedded, untestable, nonopera-
tional idea that must be replaced if we wish to understand the
patterns of history." It is noxious, he explained, because of the
social context in which progress has been used to justify racism
and the suppression of the poor and socially disadvantaged.

Since the Chicago conference, several attempts have been
made to determine objectively whether complexity does indeed
increase through evolutionary time. One involved determining
whether the structure of the spine in certain animal groups—
squirrels, ruminants, and camels—became more complex over a
period of thirty million years. Another pursued the same course
with the inner structure of the shells of ammonoids, the nautilus-
like shelled creatures that existed for 330 million years before
becoming extinct in concert with the dinosaurs. Both studies
concluded that no increase in complexity could be demonstrated. These were valiant attempts to do what few would try.
After all, the great evolutionary biologist George Gaylord Simp-

son once said, "It would be a brave anatomist who would attempt to prove that Recent man is more complicated than a Devonian ostracoderm."[13]

I'm not sure what to make of these investigations. Perhaps there was no increase in complexity in what they were measuring. Perhaps the researchers were measuring the wrong thing. Perhaps they should have been looking at some characteristics between groups, not within them. I suspect the last is valid, and agree with Simpson, who said: "An over-all trend [from simple to complex] has certainly characterized the progression of evolution *as a whole*."[14] I will stick my neck out and suggest that we look at brain evolution through the Phanerozoic, and see what change there was *as a whole*.

The most obvious change is one of size, specifically relative size. For instance, when differences in body size are taken into account, the human brain is at least a hundred times the size of any of the earliest amphibian or reptilian brains, a fact we take to be significant. In *The Descent of Man*, 1871, Darwin wrote: "No one, I presume, doubts that the large proportion which the size of man's brain bears to his body, compared to the same proportion in the gorilla or orang, is closely connected with his mental powers." Darwin undoubtedly was correct, although implicit in his statement are two assumptions that are not necessarily true and yet are widely believed. First, that bigger is smarter. And second, that vertebrate brain evolution inevitably progressed to something as powerful as the human brain.

Two decades ago Harry Jerison, of the University of California at Los Angeles, sketched the history of vertebrate brain evolution in his classic book *Evolution of the Brain and Intelligence,* and has since filled out many of the details. He developed the notion of encephalization quotient, a measure of brain size relative to body size, and tracked its change through evolutionary history. Overall, the pattern is this. With the earliest evolution of reptiles, the pattern for the group was set. Reptiles were small-brained three hundred million years ago, and they remain so today. If you plot brain size against body size (a log *x* log plot) with every reptile species you can lay your hands on, extant and extinct, you get a straight line, which speaks of a constant encephalization throughout reptilian history. The dinosaurs, many of which were much bigger bodied than living reptiles, fall squarely on this line, belying the myth that they lumbered to extinction because of limited mental powers.

When the first mammals evolved, some two hundred million years ago, these small, nocturnal creatures had achieved a leap in encephalization, being four to five times brainier than the reptilian average. When brain and body size are plotted, as previously for many species, a similar straight line is produced for the archaic mammals—it is the same slope as that for reptiles but displaced upward, reflecting the increase in encephalization. This change, incidentally, coincided with the first appearance of the neocortex, the thin layer of cells that covers the dorsal forebrain and ultimately is responsible for higher cognitive functions. The neocortex is unique to mammals, and is partly responsible for the increase in encephalization (the rest of the brain expanded, too). Even birds, which evolved shortly after mammals and achieved an equivalent level of encephalization, do not possess a neocortex.

The mammalian brain, once established in the archaic groups, remained at the same level of encephalization for at least a hundred million years. Then, with the origin of modern mammals some sixty-five million years ago, but particularly around thirty-five million years ago, another leap in encephalization occurred, another four- to fivefold increase. Ungulates and carnivores, in particular, drove this increase, with primates at the forefront, too. Prosimians, which represent the earliest form of primate and today include such species as bush babies and the galago, evolved close to the modern mammalian level early in their history. Anthropoids, which include monkeys, apes, and humans, are higher. The ratio in monkeys is two to three times the modern mammalian average, and in humans, about six times. Humans share this pinnacle with certain Cetaceans, such as dolphins.

Not all living mammals took part in the most recent jump in encephalization. Insectivores, for instance, remain close to the archaic mammalian grade, as do marsupials, such as the Virginia opossum. The fossil record of birds is poor, so the pattern of change is difficult to discern. Nevertheless, the earliest birds were close to the level of archaic mammals, while modern birds are near the average modern mammal.

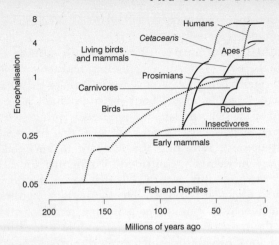

Brain size increased stepwise as new groups evolved through time.

Sketched in bold strokes as this is, the picture of vertebrate brain evolution surely misses important details. Nevertheless, several motifs stand out. The first is constancy, and the second, the apparent occasional punctuations of change. How this overall pattern is interpreted depends, perhaps, on a basic philosophy. Those who argue against any progressive, directional change in evolution emphasize the constancy within major groups. Gould is a notable proponent of this position. Those who look for fundamental drives in evolution emphasize the change. The late Allan Wilson, of the University of California at Berkeley, preferred this view.

According to Wilson, brains fueled their own further evolution. When a new behavior arises in an individual in a population, the behavior will be learned by those others in the population which are genetically predisposed to do so. The genetic predisposition to innovation and learning are therefore propelled by natural selection, forming a feedback loop that, in principle, should accelerate the process through time. Behavior, not climate change or other external force, drives evolution, said Wilson: big brains beget bigger brains, and so on.

Now, the hundred million years or so of stasis in encephalization in the archaic mammals may seem scant evidence of a positive feedback loop at work. But, said Wilson, plot the major changes through time—from reptiles, to archaic mammals, to modern mammals, to apes, to humans—and you see an escalating rate of change. "A curve of this shape implies that the process is autocatalytic," Wilson said in a lecture early in 1991, not long before he died. "That is, the brain has driven its own evolution on the lineage leading to humans."

If we think back to what Ayala said about "the ability to obtain and process information" as an adaptation that inclines to progress, then the increase in brain size between major groups through evolutionary time makes good sense. Some biologists accept the reality of an increase in complexity, but describe it as an *effect*, not a *tendency*. By effect, they mean the outcome of competition in a Red Queen situation, where an arms race, say, between predator and prey results in a swifter predator species and a more agile prey species. The improvements are a local response to a competitive situation, not a tendency of the mechanism of evolution itself. But if organisms, as complex systems, are driven to improve their ability to obtain and process information, then through the halting steps of evolutionary change, particularly in the times of major innovation, there will be a drive toward bigger brains.

I know this conclusion will not be popular among biological purists, who will accuse me of being "brain-centric." They will point to many features in the world that have had bigger impacts than the workings of collections of neurons. They are right: bacteria, for one, fungi for another. And Gould has suggested the following of a preoccupation with brains: "The not-so-hidden agenda in all this is a concern with human consciousness . . . If you believe there is an inexorable increase in brain size through evolutionary history, then human consciousness becomes predictable, not a quirky accident."[15] Nevertheless, I believe it is folly to be blind to the very real biological complexity of brains and the undeniable superiority of the human brain. I concur with Edward Wilson when he wrote, "Let us not pretend to deny in our philosophy what we know in our heart to be true."[16]

In this sense, then, humans *are* the pinnacle of evolution, the highest expression of biological computation. True, we reached this point as a result of many lucky accidents, so we cannot consider our species to have been foreordained, as Teilhard de Chardin averred. We cannot seek solace in believing that the evolution of *Homo sapiens* was inevitable. But the evolution of a mental capacity like ours, and the emergence of a degree of self-awareness like ours, probably were inevitable and predictable at some point in Earth history. We just happen to be the species in which they became manifest.

7

Endless Forms Most Beautiful

THE LAST SENTENCE of Charles Darwin's *Origin of Species* is as famous as it is lyrical:

> There is grandeur in this view of life, with its several powers, having been originally breathed into a few forms or into one; and that, whilst this planet has gone cycling on according to the fixed law of gravity, from so simple a beginning endless forms most beautiful and most wonderful have been, and are being, evolved.[1]

When the first human species arose, some five million years ago, it was but one of those "endless forms most beautiful," just as we, modern *Homo sapiens*, are today. Products of the vicissitudes of life—through an interplay of the processes of evolution and the sometimes capricious hand of extinction—we and the other species with which we share this Earth make up a global community of nearly unparalleled diversity. We saw in the previous chapter that during the last hundred million years, the engine of evolution was generating a net increase in species diversity that was unmatched in the history of multicellular life, save

for its initial explosive beginning. Interrupted briefly and spectacularly by the end-Cretaceous extinction, that increase in diversity led to a world in which more species existed in recent times
than at any other period of Earth history.

To the paleontologist, the richness of life's flow is dramatically
evident in the fossil record. Although the focus for many years of
my professional life has been on the history of the human family,
I was always acutely aware of the wider ecological context in
which it unfolded. The biota of East Africa, with which I am most
familiar, underwent great evolutionary shifts in the past fifteen
million years, driven to an important extent by great geological
change. Viewed as an unfolding drama through time, those fifteen million years witnessed a continual transformation in the
cast of characters moving through shifting scenes of complex
communities. Each of those communities was complete in itself,
but, as the fossil record shows, each had a history and, of course,
a future. Each was a transient expression of life's flow. As I will
explain shortly, I believe the paleontologist's perspective on the
nature of current biological diversity is extremely important, a
view that ecologists have only recently come to recognize.

When I became director of the Wildlife Conservation Department, I was thrust from the past very much into the present,
from a preoccupation with extinct species to a concern for species threatened with extinction. Wildlife in Kenya is tremendously varied, matching in extent the biological diversity seen in
almost any region of the world. My role as director initially was
one of implementing emergency action; namely, to put a stop to
devastating poaching, particularly of elephant and rhinoceros.
The wider biodiversity in my country was, perforce, out of my
view—for a while at least. But in moments of reflection while
away from the pressing demands of the office, such as when I
flew a small plane from Nairobi south to Tsavo National Park,
and on to the coast, west to Masai Mara, or, on all too infrequent
occasions, north to Lake Turkana, I was reminded of the richness
of life below.

Ecologists speak of three measures of biological diversity. The
first, alpha diversity, reflects the number of species within an
ecological community. Beta diversity compares species composition in neighboring communities that differ in certain physical
characteristics, such as elevation. Gamma diversity, the third

measure, encompasses communities over a wider geographical range and so may include regions with similar habitats that are separated by many miles. The flight from Nairobi to Lake Turkana is a study in these three measures.

Nairobi lies at five thousand feet, resting as it does on the vast geological dome that fifteen million years ago heaved the continental crust up from near sea level to more than nine thousand feet at its highest point. Flying north out of Nairobi, I have to negotiate the shoulder of the Rift Valley, which rises some four thousand feet above the city. The terrain, extremely fertile, is rich, red volcanic soil supporting a patchwork of tea and coffee plantations and small towns. Breasting the shoulder of the rift is always dramatic, with arresting terrain in all directions and protean cloud formations above and below—and it's always a relief when I clear the hazardous ridge.

To the west, the walls of the valley drop away precipitously, creating a contrast between verdant highlands and parched valley bottom. The Aberdare Mountains, to the east of the flight path, are nourished by abundant moisture, and support a wonderful diversity of wildlife, including elegant black-and-white colobus monkeys and even leopards. There used to be tens of thousands of elephants, too, but no longer, as poaching and the spread of agriculture in this fertile terrain decimated their numbers; about five thousand live there now. Beyond the Aberdares to the east is Mount Kenya, rising to a snow-covered peak at just over seventeen thousand feet. Tremendous contrasts abound in this single vista: mountaintop glaciers, alpine meadows, and thick temperate forest on Mount Kenya; lush, humid forest on the lower slopes of the Aberdares; arid desert on the rift floor; and a complex and graded mosaic of vegetation linking them. I know that in each of these habitats, aside from the frigid glaciers, the diversity of animal and plant life is great; that's high alpha diversity. And I know that if I were to go from the lower to the upper slopes of Mount Kenya, I would be visiting very different biological worlds, from subtropical to alpine, within half a day's hike; that's high beta diversity.

The contrast among the ecological communities I fly over in the almost three-hour trip from Nairobi to the east side of Lake Turkana is breathtaking, covering, as it does, all I've mentioned so far, and more. I pass over the edge of the Laikipia Plateau, after which the last hour and a half of the flight gives views of lava flows and craters, dry lake beds, shadows of evanescent water courses in the dry earth, and finally the sandstone stretches that constitute the eastern shore of the giant lake. There's less alpha diversity where we land, of course; low rainfall and high temperatures determine that. Even so, there's more life in the region than is evident at first glance of most visitors, and sufficient to support herds of wildebeest, topi, prides of lion, and even leopards. The trip has given a glimpse of Kenya's gamma diversity, comparing communities over a considerable geographical range; again, it's high, and the importance of topological variation in generating that diversity is obvious. The rich mosaic of high and low elevation, forming a myriad of habitats that differ in daily temperature range, availability of moisture, and many microclimatic features, supports—indeed, generates—a rich mosaic of ecological communities.

Were I to fly west from Nairobi instead of north, crossing the Rift Valley into Uganda and beyond, before very long my view would be of unrelenting green, a carpet of tropical forest from horizon to horizon. This is the home of the African great apes. Tropical forests host great diversity, it is true, but without the topological variations we see in the Rift Valley, they lack contrast, an important element in gamma diversity; there are no herds of plains animals here, no lizards adapted to arid desert life, no alpine flowers. In addition to much else, the variation in topology east of the Rift Valley, with its mosaic of habitats, provided the impetus for the early evolution of the human family.

I've talked about my own experience, both in paleontology and in wildlife protection in Kenya, as a way of introducing the notion of biodiversity in the present world, the theme of this chapter. I'll now extend our view beyond Kenya, beyond Africa, to encompass the globe, and in so doing I will address two questions, both of them central to modern evolutionary ecology. The first concerns the shape of biodiversity and the processes that generate it, on both local and global scales. We'll see that, while the shape may be described with some confidence, being simply what is evident before our eyes, the processes underlying its origin are not at all obvious. The second question, that of how many species constitute the global diversity of which we are a part, may be thought to be readily answered. It's not. I'll explain why this is so.

There are many components to the overall shape of biodiversity, but I'll concentrate on two of the most important. First is the global distribution of life, or where most species are to be found. The second compares the diversity between life in the oceans and life on land. In some fundamental manner, these are linked to each other, through the engine of evolution and dynamics of ecosystems.

The most striking pattern in biological diversity throughout the world is its unequal distribution. Very simply, species diversity is highest around the equator and diminishes steadily at ever higher latitudes; that is, as one moves north or south toward the poles. Anyone who lives in North America or Europe and has visited Kenya, for instance, sees this contrast immediately. It is evident in the realm of animal life, not just in the great panoramas of migrating herds, and the lions, leopards, and cheetahs that attract so much of visitors' attention, but also in the fabulously rich bird life and, of course, the swarming insect world. It is evident, too, in the realm of plant life, particularly in moist forests, with an abundant mixture of tree species and epiphytes that grow on them, and the microepiphytes that grow on *them*. And if we were to look at a microlevel—at fungi and bacteria— we would see the same super-richness in Kenyan life forms, which far outstrips that in most visitors' home countries.

Termed the "latitudinal species-diversity gradient," this bold signature of nature has been known to biologists for many years. It has inspired countless hypotheses in attempts at its explanation and, in so doing, essentially gave birth to theoretical ecology. It also has considerable implications for conservation biology: a square mile of habitat destroyed in the tropics potentially endangers at least ten times as many species as a similar area loss in temperate regions. Tropical rainforests are especially rich in biodiversity: they cover one sixteenth of the world's land surface, yet are home to more than half its species. For this reason, the relentless destruction of these forests is of pressing concern.

This focus on rainforests should, however, not be at the expense of other tropical habitats, as a recent study showed. In a survey of almost a thousand mammalian species in South America, the University of Oklahoma biologist Michael Mares found that, for certain types of species at least, the richest diversity was to be found, surprisingly, in dry regions (such as the llanos of Venezuela and Colombia, the cerrado grasslands and caatinga scrublands of Brazil, the chacoan thorn forest, and the Argentine pampas). "Drylands are often considered areas of low diversity," he stated in his report in *Science*, "but for mammals they are the most species-rich area on the continent."[2] The discovery by Mares does not negate the tropical rainforests as valued regions of high biodiversity, but widens our appreciation of where diversity is to be found. As the University of Tennessee ecologists Stuart Pimm and John Gittleman wrote, in a commentary on Mares's paper, "We know too little about where diversity is, why it is there, and what it will become."[3]

Before I address some of the suggested reasons for species abundance in the tropics, I will give some brief examples of the steepness of the gradient. If, as an ant lover, you made a trek from Alaska to Brazil, counting the number of species in each region as you went, you would start with just three at the beginning of your journey and end with some 222 at your destination. That is a difference of almost two orders of magnitude. The best known ant lover, of course, is the Harvard biologist Edward Wilson, who told a scientific gathering on biodiversity, held in Washington in 1987, that "from a single tree in Peru I identified forty-three species of ant, approximately the same number as in all the British Isles."

The story is the same for bird and tree enthusiasts. For instance, the Harvard biologist Peter Ashton counted the number of tree species in twenty-five acres of forest in Borneo, and came up with seven hundred. The same figure describes the number of tree species within *all* of North America. For land birds, the species count from Alaska to the American tropics is also dramatic, going from twenty to six hundred. George Stevens, an ecologist who has spent time in both these regions of the world, summed up the difference as it appears to the eye: "When you travel around Alaska you are struck by how very biologically dull it is. Yes, the geology is fascinating, but there is a sameness about the fauna and flora, no matter where you are. But in Costa Rica, even with the slightest differences in terrain, you see great differences in habitat."[4]

Again and again, the species-gradient pattern is boldly drawn in the terrestrial realm. Until recently, there were mere hints of how it may play out in the marine realm, which is much less accessible to the ecologist's eye, particularly with respect to life in the deep oceans. Once considered something of a biological desert, the deep oceans are now known to harbor a tremendous range of life forms. And, since the first discoveries, in the 1970s, of creatures sustained by energy from the thermal vents associated with volcanic action of tectonics, some of these forms are known to be of the most bizarre kind.

The newly emerging picture of life under the waves, therefore, is strikingly like that on land, as confirmed in a major study by a team of researchers from the United States, Scotland, and Australia, at the end of 1993. The highest diversity is concentrated around the equator, with species richness falling off as one looks under seas at higher latitudes. "Latitudinal gradients of diversity were unexpected," the researchers wrote, "because it was assumed that the environmental gradients that cause large-scale patterns in surface environments could not affect communities living at great depths."[5] The deep sea has always been viewed as relentlessly monotonous, ecologically speaking, whatever the latitude, hence the assumption that there would be little difference in diversity between different latitudes. The researchers correctly point out that their discovery has implications for conservation similar to those of habitat destruction in the terrestrial realm. For instance, mining, petroleum exploration, and waste disposal operations will have different scales of impact on biological diversity when they are carried out in tropical waters rather than in Arctic waters.

The statement that the species gradient in the deep sea was "unexpected" brings us to the issue of what causes the global pattern, and it reveals at least one apparently reasonable assumption: the gradient in species richness relates directly to gradients in important physical elements on land, such as temperature and light. Intuitively, we feel that habitats on land and in the deep sea must be very different, and they are. Many deep-sea habitats are pitch black, and fluctuations in temperature are buffered. Not so on land. Clearly, some fundamental mechanism is operating here, expressing itself in very different environments. The problem with discerning what mechanism drives this pattern is not a lack of hypotheses, but a plethora of them. Over the years, many explanations have been offered, often directly contradicting one another. Such contradictions are a salutary indication of how far we are from being confident about a resolution of the issue. I'll describe some of the hypotheses.

One long-time favorite is the so-called time theory, which argues that conditions in the tropics have prevailed longer than those in temperate regions. The cause was periodic glaciation, which affects temperate regions disproportionately. Biodiversity has therefore had longer to accumulate in the tropics, or so it is suggested. However, there are regions of the world that were relatively unaffected by glaciation, including some of the more northerly stretches of terrain, and these do not show the higher species richness the theory would predict. The time theory clearly doesn't hold. Neither does the "productivity hypothesis," which has also been popular for a long time.

The tropics do seem blessed by nature: they are bathed in balmy temperatures, abundant light, and, in many areas, a plentiful supply of water. We know from our daily experience that plants thrive in such conditions. And, as animals depend on plants, a high diversity—of plants and animals—can therefore be supported. This is true, but there is a logical jump in assuming that high productivity—that is, a high biomass—necessarily generates high species diversity. Why should there be an abundance of species sharing this nurturing environment rather than a few species luxuriating in their own abundance? (Coniferous forests of the north support huge biomass, but few species; grasslands may be low in biomass but high in species number.) The weakness of the theory is that there is no obvious link between high productivity and the generation of many species.

Notice that while the time hypothesis does not seek special evolutionary properties of the tropics, merely additional time for the accumulation of species, the productivity hypothesis does adduce such specialness. It implies that there is something about the tropics that encourages a higher rate of speciation than occurs in temperate regions. When you walk through bushland savannah or dense, moist forest, you feel yourself amid a kaleidoscope of species, a richness of life on all levels, like fractal patterns, which seems to speak of evolutionary creativity. The assumption that the tropics nurture the flow of life, producing "endless forms most beautiful" at an elevated rate, *seems* right. But there is another explanation: the tropics may simply be a more forgiving environment, driving species into extinction less frequently than happens at higher latitudes. Intuitively, this is attractive, too; animals and plants have to survive through harsh winters at high latitudes, when mortality can be high. Populations made vulnerable through locally reduced numbers might succumb to extinction in a particularly severe year. The struggle for existence does appear to be greater at higher latitudes, with their wide seasonal fluctuations.

Until recently, it wasn't possible to say whether the accumulation of tropical diversity was the result of extraordinary evolutionary innovation or of an ameliorated rate of extinction. A couple of years ago, David Jablonski addressed the ecologists' question, through recourse to the fossil record. He reasoned that if the record shows that the first appearance of new species occurred more often in tropical regions than in temperate ones, then the question is answered. He chose the record of marine invertebrate species since the beginning of the Mesozoic, some 225 million years ago, to address the question, and saw a clear signal of higher numbers of first appearances in the tropics. "This provides direct evidence that tropical regions have been a major source of evolutionary novelty," he reported in *Nature* in July 1993, "and not simply a refuge that accumulated diversity owing to low extinction rates."[6] This important result focused biologists' questions more clearly: whatever is special about the tropics, it *does* promote evolutionary innovation.

These days biologists agree that one of the most important ways in which new species arise is through what they call allopatric speciation. This simply means that populations of an existing species become isolated from one another by some means, and over a relatively short period of time (a few thousand years) accumulate sufficient genetic differences and adaptations to become independent, daughter species. (This was the core of Elisabeth Vrba's turnover-pulse hypothesis I described earlier.) How might environmental conditions in the tropics promote allopatric speciation? Two suggestions have been made, and I'll describe both.

Before I do, however, I should point out that it would be naïve to imagine the tropics as a belt of uniformly fecund habitats, each pumping out evolutionary innovation at a rate higher than in temperate latitudes. Biologists have come to realize that habitats differ in their potential for supporting populations of the same species, so that, for instance, in one habitat the population may thrive and expand while in an adjacent one it withers away. However, because animals and plants (via seed dispersal) are mobile, individuals from the first habitat may continually migrate to the second, giving rather uniform populations to both. The two habitats are referred to, respectively, as sources and sinks, for obvious reasons. Source and sink habitats are likely to exist for species production, too, so, for instance, an apparently uniform swath of rainforest may in fact be a mosaic of habitats that differ in their potential to promote speciation.

For many biologists, this potential derives from the environmental stability of tropical habitats. "When the climate is more stable, the supply of resources is likely to be more stable," explains Wallace Arthur, an ecologist at Sunderland Polytechnic in England. "Consequently species can afford to be more choosy about their food—they can have narrow food niches—and still survive."[7] The narrowness of food niches encourages these specialist species to remain in limited geographical ranges, where the required food resources are. Not only does this mean that many more species are likely to inhabit a given area, but it also promotes speciation. The equable nature of tropical climes permits small populations to survive, even at the edge of their range, suggests Arthur, whereas in the harsher northern latitudes such populations are more likely to disappear; new species are therefore less likely to originate there. In this scenario, stability is viewed as the engine of evolution and also as providing an environment in which many species can coexist. Is the scenario correct?

Not according to a more recently developed hypothesis, which identifies *instability* as the engine of evolution. During the 1970s and early 1980s, evolutionary ecologists were engaged in a vigorous debate over the forces that shape community structure. At the beginning of the debate, the conventional wisdom was that competition between species determined the composition of communities, particularly which species could coexist with others. Within a decade, competition, while not dismissed as a factor in shaping community structure, had been very much played down. Joseph Connell, an ecologist at the University of California

at Santa Cruz, was influential in promulgating this shift. He re-
calls visiting Australia to study reef communities, as a convinced
competitionist, only to have his view changed by nature: a hurricane promptly swept through his study site and wiped out large
proportions of the community. For Connell, it was a salutary
demonstration that forces other than competition are important
in the world of nature. While a hurricane may be a rather dramatic and unusual form of environmental perturbation, lesser
disturbances are more common. They are also more creative,
albeit in an indirect way.

Communities that are immune to disturbance may soon become dominated by a few species; stress the community, however, by, say, making gaps in forest cover, and other species have
an opportunity to be part of the community. By extrapolation,
repeated stress of this nature may promote the evolution of new
species. This may seem counterintuitive. It is easier to imagine a
benign, stable environment as the cradle of evolution, nurturing
emergent species in their most vulnerable state. But the reality
increasingly seems to encompass stress and instability as the midwife of evolution.

The species-rich rainforests of the Amazon, for example, have
been subject to tremendous perturbation, particularly during the
end of the last Ice Age, ten thousand years ago, and the beginning of the more equable recent times. There was no continentwide carpet of forest during the earlier, frigid times. Instead, fragments of forest existed in favorable habitats,
microclimates that protected species adapted to warm climates.
There is a long-running, continuing, and unresolved debate over
how such fragmentation promotes speciation. One notion is that
in the forest fragments, or refugia, populations of a species were
isolated, and therefore drifted away genetically from other, similar populations in other refugia. This fits the hypothesis of allopatric speciation very well. But convincing evidence of such
refugia has been hard to find. Perhaps the simple agent of environmental perturbation was the engine of evolution, a creative
environment balanced between complete stability and complete
instability—or chaos. (In the Amazon rainforests, the great Ma-
yan civilization added to this perturbation by leveling much of
the forest in certain areas.)

Nature's creativity in the region between stability and chaos is also to be seen in the oceans, where a curious pattern exists. The deep sea is where the great biodiversity of the marine world resides. Nearshore communities, by contrast, are relatively barren. New species arise in both these regions, more prolifically in the deep sea. And yet the greatest evolutionary novelties—new species with entirely new adaptations rather than mere variation on existing themes—occur more often in nearshore communities. The phenomenon appears to be real; that is, the result of greater innovation rather than the preferential survival of evolutionary oddballs in nearshore as opposed to offshore communities. The reason for it, however, remains a mystery. What can be said of nearshore communities is that, because of wave action, they are much more disturbed than those in the deep sea. Stress, it appears, encourages evolutionary innovation.

If disturbance is the midwife of evolution, why would there be a difference between tropical and temperate regions? One reason, proposed recently by George Stevens, is that while tropical species often are specialists, adapted to narrow environmental ranges, temperate species by their nature must be more broadly adapted, tolerating the fluctuating temperatures and light flux of the seasons. Tropical species are therefore more vulnerable to disturbance, which also provides the evolutionary opportunity for speciation; temperate species are much more tolerant, and as a corollary are pushed to the edge of evolutionary creativity less frequently.

What of the deep sea, where the same latitudinal-species gradient is generated as on dry land? There is a prevailing view of the deep sea as a region of endless sameness, a place of little change, isolated as it is from the vicissitudes of the environment. If this is true, then the disturbance hypothesis falls. But as the ecologists John Gage and Robert May suggest in their commentary on the discovery of the deep-sea species gradient: "Maybe the deep-ocean floors are not so globally uniform after all."[8] Perhaps so. After all, it wasn't very long ago that the deep sea was regarded as a biological desert, and yet it is now known to boast a level of biological diversity that, in places, rivals that of tropical rainforests.

We have seen, therefore, that the pattern of global diversity could hardly be more striking and obvious. And we have also seen that in generating "endless forms most beautiful," the pro-

cess of evolution works in ways that still cheat biologists' understanding.

The second most important pattern in global biodiversity comes
from a comparison of terrestrial and marine realms. Of the total
number of species recorded to date, fewer than 15 percent live in
the oceans, and mostly on or near the deep-sea bed; the rest live
on land. As the oceans cover almost three quarters of the globe's
surface, this imbalance is especially dramatic. On the face of it,
the terrestrial realm appears to support far greater diversity than
the marine realm. There is, however, a paradox, that relates to
the level in life's hierarchy at which one compares these two
realms.

I've been talking here about species, which is the lowest level
of the biological hierarchy. If, now, we do a comparison of diversity in the marine and terrestrial realms at the level of the phylum, we see a picture very different from that of species diversity.
Of the thirty-three animal phyla, representatives of thirty-two live
in the sea and only twelve on land. Moreover, some 64 percent of
phyla occur only in the marine realm, as against a mere 3 percent that are exclusively terrestrial. (The balance are to be found
on both land and in the oceans.) On this measure, the seas support a far greater diversity of life forms than do terrestrial habitats. In other words, in the sea, there are to be found many
themes, with few variations on each, while on land, there are
many more variations on fewer themes.

As with the latitudinal-species gradient, explanations for the
land-sea differences in diversity at the phylum and species levels
are legion, a sure indication of uncertainty. Even at a basic level
it is sometimes difficult to answer the simplest of questions, such
as why terrestrial life is dominated by insect species while very
few are to be found in the sea.

The most obvious reason that there are more phyla in the
marine realm is that multicellular life began there, in the Cambrian explosion. All existing phyla evolved at this event or soon
afterward, long before the first organisms ventured onto dry
land. All existing phyla would therefore have had an opportunity
to leave descendants in the oceans, whereas only those which
evolved a terrestrial adaptation had a similar opportunity on
land. This does not address why there are far fewer variations of

each body plan in the marine realm than on terra firma—after all, life in the seas had a hundred-million-year head start on life on land. There must be something different about the two environments that gives more power to the engine of evolution in one place.

If the deep oceans were the vast stretches of sameness that had until recently been assumed, then a lower species diversity would be explicable. Heterogeneity in the environment—in topography and climate—and its change through relatively short periods of time promote the evolution of species. But, different in several important ways though the deep sea environment may be, it is probably much more heterogeneous than has been supposed.

The great forests of the tropics and, until recently, temperate zones offer a spatial complexity that, as far as we know, is not common in the sea (except in coral reefs, which support high species diversity). While accepting that this may play a part, Robert May, who conducted a study of the problem, remains skeptical: "I have difficulty . . . in accepting that this could explain the 85:15 land-sea difference in species numbers."[9] He also points out a curious difference between tropical forests and coral reefs, which are often considered equivalent in their different realms. "The coral reef teems with conspicuous animal life . . . by contrast, in the tropical forest vertebrates are rarely glimpsed, and even invertebrates are not all that conspicuous."[10] Why there is this difference, no one can say.

Many other hypotheses have been proffered to explain the land-sea difference in species richness; the most promising one, perhaps, has to do with species size and range. Smaller on average than terrestrial species, species in the marine realm also have a wider geographic range. And a wider geographic range translates—usually—into fewer species overall. To those with an evolutionary ecology perspective, this explanation sounds promising, but it is, admittedly, speculative.

From patterns in global diversity, I'll turn now to the composition of that diversity, and ask: How many species are there in the world today?

To this simple question Robert May has a simple answer: "We do not know to within an order of magnitude how many species we share the globe with."[11] Most estimates fall between five and fifty million, with some as high as a hundred million. The reason for this wild uncertainty? Few biologists have even tried to find an answer, and those who have are daunted by the difficulties of reaching a reliable one.

It is indeed a remarkable fact that we in this modern world, obsessed with measuring things, are so imprecise about the stuff of nature, to which we are intimately related and upon which we ultimately depend. We have a good estimate of how many stars there are in our galaxy, the Milky Way: some hundred billion. We know how many nucleotide bases constitute the human genetic blueprint: three billion. And we can calculate to within a few hours when a comet will collide with Jupiter, as it did at 4 P.M. (U.S. Eastern Daylight Time) on 16 July 1994. Yet we cannot put a secure number on current species diversity. It is not through lack of knowing how to obtain it, but through lack of commitment. Governments have spent hundreds of millions of dollars in the systematic study of the stars but only a tiny fraction of that sum on a systematic study of nature here on Earth.

The search for order in nature essentially began with Aristotle, but almost two millennia were to pass before natural history became a respected discipline in Western science. Classification of plants and animals emerged as a principal endeavor of the new science, initially in the context of demonstrating the products of God's handiwork. The modern system of classification was established by Linnaeus in his *Systema naturae* in the mid-eighteenth century. He recorded some nine thousand species of plants and animals, using a system that revealed their relation one within another (in a sense of creation, not of history, of course). This groundbreaking but modest first venture into a systematic study of the world of nature came a full century after Isaac Newton had developed the laws of gravity, upon which the calculations for the recent comet-Jupiter collision were based.

Since Linnaeus's time, the number of recorded species has increased substantially, of course, and currently stands at about 1.4 million. I say "about," because there is no central repository for all species descriptions, so the figure is necessarily an estimate. This is especially ironic; there *is* a central repository for the DNA sequences that are produced in molecular biology laboratories around the world, but not for the organisms from which the genetic material was obtained.

Eighty-five percent of recorded species live in the terrestrial realm, and the majority of these, some 850,000, are arthropods (that is, insects, spiders, and crustaceans). Most of the arthropod species are insects, and almost half of these are beetles, a fact that is said to have inspired a famous epigram from the British biologist J.B.S. Haldane. On being asked, one day, by some clerical gentlemen what his study of the natural world had revealed to him about God, Haldane is said to have replied that it indicated that He had "an inordinate fondness of beetles." Probably apocryphal, the story is nevertheless apposite, and, if recent studies are reliable, may even underestimate reality.

Of the 300,000 known species of plant, most are flowering plants. (The separate abundances of insects and flowering plants is surely no accident, for they have progressed in coevolution with each other these past several hundred million years.) Some sixty-nine thousand species of fungi have been identified and about the same number of single-celled organisms. Within this last figure, only five thousand species of bacteria have been described. The realm that occupies most of our attention, that of vertebrates, boasts a total of forty thousand, of which four thousand are species of mammal, nine thousand are birds, and the rest reptiles, amphibians, and fish. In case anyone might wish to imagine that the four thousand species of mammal measure up tolerably well to the count for bacteria, they should pause before swelling with pride. Recently, a Norwegian research group analyzed the bacterial content of one gram of soil from a beech forest and a similar quantity of sediment off the Norwegian coast. In each case the group identified as many as five thousand species, with no overlap between the samples. Evidently, the five thousand bacterial species recorded so far are but a tiny sample of reality.

In addition to being a substantial underestimate of the actual numbers of species in the world, the list of those known is biased in several ways. First, it reflects the not unnatural human interest in furred and feathered creatures. Far more taxonomists work on birds and mammals, say, than on insects or nematodes or bacteria. As a result, although new species of birds and mammals are discovered every year, they are few, and the ultimate total may not be significantly different from the current count. This is not the case for the rest—indeed, most—of nature, as the example of

Number of Living Species of All Kinds of Organisms Currently Known
(According to Major Group)

ALL ORGANISMS: TOTAL SPECIES, 1,413,000

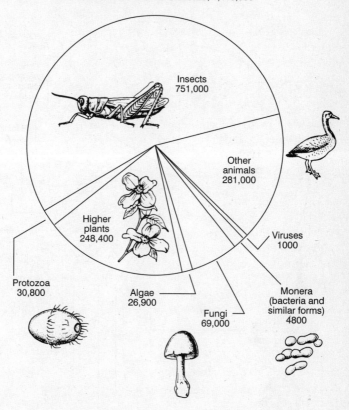

Insects
751,000

Other
animals
281,000

Higher
plants
248,400

Viruses
1000

Protozoa
30,800

Algae
26,900

Fungi
69,000

Monera
(bacteria and
similar forms)
4800

Insects and higher plants dominate the diversity of living organisms known to date, but vast arrays of species remain to be discovered in the bacteria, fungi, and other poorly studied groups. The grand total for all life falls somewhere between 10 and 100 million species.

(Reprinted by permission of the publishers from The Diversity of Life *by Edward O. Wilson, Cambridge, Mass.: The Belknap Press of Harvard University Press, Copyright © 1992 by Edward O. Wilson.)*

Number of Living Animal Species Currently Known

(According to Major Group)

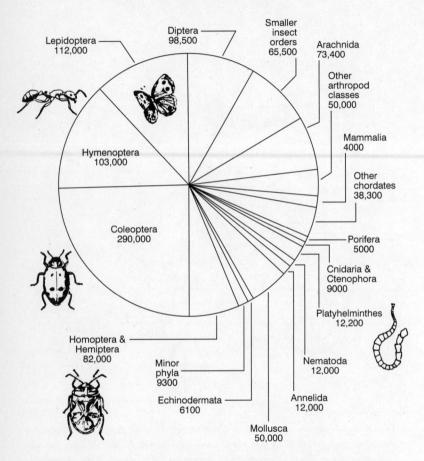

ANIMALS: TOTAL SPECIES, 1,032,000

Diptera 98,500

Smaller insect orders 65,500

Lepidoptera 112,000

Arachnida 73,400

Other arthropod classes 50,000

Hymenoptera 103,000

Mammalia 4000

Other chordates 38,300

Coleoptera 290,000

Porifera 5000

Cnidaria & Ctenophora 9000

Platyhelminthes 12,200

Homoptera & Hemiptera 82,000

Nematoda 12,000

Minor phyla 9300

Annelida 12,000

Echinodermata 6100

Mollusca 50,000

Among animals known to science, the insects are overwhelming in number. Because of this imbalance, most animal species live on the land; but most phyla (Echinodermata, etc.), the highest units of classification, are found in the sea.

(Reprinted by permission of the publishers from The Diversity of Life *by Edward O. Wilson, Cambridge, Mass.: The Belknap Press of Harvard University Press, Copyright © 1992 by Edward O. Wilson.)*

the bacteria demonstrates. A second bias is a concentration on northern temperate regions of the world, where most taxonomists (those who classify organisms) work. The bulk of species live concentrated around the tropics, and yet for every species known in the tropics, two are known in the higher latitudes in the north.

One of the first attempts to pull together a scientific estimate of the number of insect species was made by the British ecologist Carrington Williams, which he published in a 1964 book called *Patterns in the Balance of Nature*. He arrived at a figure of three million, through a combination of local sampling and extrapolation. During the next two decades, field biologists, usually working independently, gathered information in many different habitats, some of them novel, such as the deep sea floor. As a result, the total species count edged up to at least ten million.

Then, in what was one of the more dramatic events in systematic biology, Terry Erwin, of the Smithsonian Institution, announced in 1982 that there were probably at least thirty million insect species alone, mostly in the canopy of tropical rainforests. Erwin obtained his estimate by counting the insect population in a large sample of trees in the Panamanian rainforest, a task he achieved by releasing insecticide high in the canopy, and collecting the corpses as they fell to the ground. A daring piece of work, but, important though it was for science, it went largely unnoticed. As Edward Wilson commented a few years later, "If astronomers were to discover a new planet beyond Pluto, the news would make front pages around the world. Not so for the discovery that the living world is richer than earlier suspected, a fact of much greater import to humanity."[12]

Since Erwin's announcement, interest has been gathering in the unanswerable question. Again, it was not a coordinated effort; it came from biologists working independently in their own fields of research. For example, David Hawksworth, of the International Mycological Institute in Kew, England, argues that the current estimate of some 69,000 species of fungi is at least a twentyfold underestimate. He showed that for every species of vascular plant in European habitats, there are about six species of fungi; and as there are about 300,000 known species of such plants, there are probably close to 1.8 million species of fungi. If more species of plant are discovered, then this greatly elevated

Nest of the water hen—Fulica chloropus *(Linnaeus).* —*From the Rouen Museum.*

Return of ants after a battle.

The bird-eating spider (Mygale avicularia) *killing a hummingbird. —From Sybille de Merian.*

figure will rise still further. A similar picture holds for nematodes, suggests Peter Hammond, of the Natural History Museum in London. Some 15,000 species of these tiny worms are known. Ubiquitous parasites on animals and plants, or living free in marine and fresh water, nematode species, Hammond calculates, may number at least 300,000 worldwide.

Calosoma (Calosoma inquisitor) *pursuing a bombardier* (Brachinus crepitans), *which is fighting in retreat.*

These and other estimates bring the total number of species to close to fifty million, which even then may be too low. In what he describes as a "flippant but not entirely unreasonable" scenario, Robert May calculates a total of a hundred million species. He bases this on the notion that "every species of arthropod and vascular plant (which together account for the vast majority of recorded species) has at least one parasitic nematode, one protozoan, one bacterium, and one virus specialized to it."[13] It should be obvious by now that, whether the total species count is thirty, fifty, or a hundred million, most of life is tropical and inconspicuous. The world of large vertebrates and large plants that is our daily experience is but a fraction of the diversity of life. We see the *shape* of life's diversity, with relatively few large organisms and relatively many small ones, and understand that it has to do in part with the flow of energy through ecological communities. But we are unable from first principles to predict the *extent* of the diversity; there is no theoretical basis in ecology and evolutionary biology for saying that the Earth in its present configuration of continents should support one, ten, thirty, fifty, or a hundred million species.

There is uncertainty in all the figures I've been giving for species numbers, because each is based on an extrapolation from some sort of field measure. Inevitably, some calculations will go

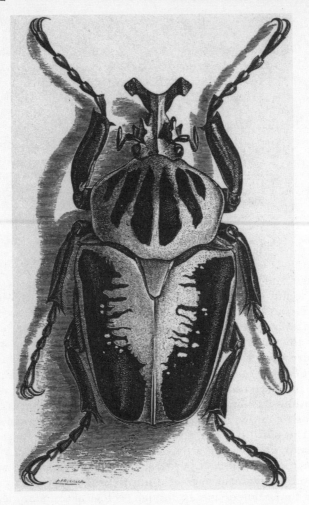

Goliath of Drury—Golianthus giganteus.

awry when ratios between numbers of species (such as fungi to plants) differ in different parts of the world. Nevertheless, it is the most feasible way of arriving at a species count. It has taken biologists some 230 years to identify and describe three quarters of a million insects; if there are indeed at least thirty million, as Erwin estimates, then, working as they have in the past, insect taxonomists have ten thousand years of employment ahead of them. Ghilean Prance, director of the Botanical Gardens in Kew, estimates that a complete list of plants in the Americas would occupy taxonomists for four centuries, again working at historical rates.

These insect and plant taxonomists were doing more than just counting species; they were describing them, too. Each catalogue entry is an encapsulation of a unique form of life, a heritage of hundreds of millions of years of evolution, of which we are but one part. But the number of entries is woefully small, and the challenge to increase it is enormous, given the resources Western science has devoted to the task so far. It is a sad commentary on the value placed on the great diversity of life here on Earth that, as May said, "we do not know to within an order of magnitude how many species we share the globe with."

Should we, as Edward Wilson has urged, "aim at nothing less than a full count, a complete catalog of life on Earth"?[14] Such a venture would be costly, to be sure, but much less so than sequencing the human genome, which consumes $150 million a year, or building a space station, with a total price tag of some $30 billion. As director of the Wildlife Conservation Department, which became the Kenya Wildlife Service almost a year after I took over, I experienced, day by day, the cost of trying to preserve species, usually the most conspicuous ones. And I can see that the expenditure of, say, $100 million a year would be welcome in many countries that are struggling to preserve wildlife in the face of continued growth of human populations. But Wilson's ambition is worthy: it is worthy of science and, more fundamental, it is worthy of humanity. As the pinnacle of the evolutionary process, a sentient species, we have a moral duty to know as much as can be known about the "endless forms most beautiful" with which we share this Earth.

8

Value in Diversity

THESE DAYS, whenever ecologists talk about biological diversity, they usually feel obliged to justify its value. A quarter of a century ago, no such obligation was felt, for few people bothered to talk about diversity at all. The question of its value therefore did not arise. Earlier still, around the turn of the century, the value of diversity wasn't an issue either, but for different reasons. The kaleidoscope of species that constitute the diversity of life was recognized as an integral part of life. Naturalists marveled at the creativity it revealed, whether they saw it as the product of God's divine hand or of the process of evolution. Now, however, discussion of the value of biodiversity has become "a cottage industry," to use a phrase of the Rutgers University biologist David Ehrenfeld. There are, he says, "dozens of us sitting at home at our word processors churning out economic, philosophical, and scientific reasons for or against keeping diversity."[1]

The reasons for this shift in attitude are several. Important among them is the greater understanding of the complexity of the natural world that has emerged from the maturation of the science of evolutionary ecology over the past decade or so. Rooted in the keen observations of many generations of naturalists, ecology began to be established as a science only in the past half century, with pioneers like Charles Elton, G. Evelyn Hutchinson, and Robert MacArthur. The marriage between evolutionary biology and ecology is younger still, and one of its main products is our ability to view the current world in the context of continuous change. We see individual organisms not just as isolated, natural phenomena that demand our appreciation, but as components of ecosystems that evolve together. Biodiversity, therefore, is an expression of that collective change; the mechanisms that underlie its generation still hold many secrets, but we understand the phenomenon a great deal more than we did. We stand in wonder at its intricacies and its interconnectedness. In a sense, through a greater mechanistic understanding of diversity, we have regained the nineteenth-century naturalists' awe of nature. And we value what we see, but in the context of science.

The more pressing reason that ecologists now so frequently speak of the value of biological diversity, however, is the recent recognition that it is under an accelerating threat of destruction. Ecologists want to know not only the consequence of the loss, which is the converse of stating its value, but also wish to marshal arguments for halting that loss. The specter of the extinction of species—largely through the destruction of habitat in the face of industrial and agricultural expansion, both of which are aspects of continued population growth—will become ever more clear as we move toward the last sections of the book. In this chapter, I will address the issue of assessing the value—in its several forms —of biodiversity, as prompted by the recognition of its growing predicament. As we've seen, there are perhaps fifty million species on the Earth today. What does it matter if we lose some?

The naturalists of the nineteenth century would have replied that it matters because each species is part of the whole, and we should value it because of its contribution to the whole. Many modern ecologists feel this way, too, but the tenor of the debate over value has been cast very differently. Because the threat to diversity comes from the world of material resources, the ecologists' response has been increasingly couched in material terms.

Logging and mining companies know how many dollars they will reap when, in the process of destroying natural habitat, they harvest trees or recoup minerals from the ground. The same applies to ranchers who turn forest into temporary pasture land. Much of the attention of the cottage industry to which David Ehrenfeld referred centers on the dollar value of the habitats that would vanish if they were exploited for commercial gain. If saving species and ecosystems can be shown to promise more value than destroying them, then there is a strong reason for doing so, or so it is argued. "People are afraid that if they do not express their fears and concerns in this language they will be laughed at," comments Ehrenfeld.[2] As a result, ecologists have largely allowed economists to set the terms of the debate over the value of biodiversity. The danger is that, having accepted the invitation to enter the lion's den, they are likely to end up as the lion's dinner.

The problem of placing a value on biological diversity is extremely complicated, and, as I learned daily from my experience as director of the Kenya Wildlife Service, economics *are* important. For instance, the principal preoccupation of local people is survival. They therefore must see and experience economic benefit from the maintenance of parks where wildlife thrives; otherwise, how can they be expected not to want to use the land for pastoral or agricultural purposes? There are many ways in which this very real concern may be addressed, and it is a matter over which there is much controversy. But economics cannot—and must not—exclusively set the limits of the debate. I want to explore these limits. Some of the ground I will tread is familiar territory to those who have made biodiversity a concern over recent years. Other considerations are newly emerging, and some of them relate closely to my own passion of three decades, the place of *Homo sapiens* in the world of nature. We are a product of evolution that includes us as but one species among many in the global complexity that is the biosphere.

I identify three principal ways in which we should value biodiversity. The first falls squarely into the economic realm, to which I have alluded. It includes the tangible benefits we can extract from our environment, such as food, raw materials, and medicines. The second benefit is less tangible in the traditional sense, but no less important; this is the maintenance of the physical environment, in its circulation of gases, chemicals, and moisture.

It relates to the continued health of the global environment, upon which we and our fellow species depend for survival. The third realm of value is less tangible still; it is the esthetic pleasure individual humans derive from their experience of the diversity of life around them. As others have argued, I believe this goes beyond a merely abstract experience and, instead, taps deep into what it is to be human. An appreciation of—and psychological dependence on—biodiversity is part of the biologically built psyche of *Homo sapiens,* a product of a long evolutionary history. If biodiversity is depauperated, by artificial or natural agency, so too is a fundamental component of human existence.

The issue of biodiversity and economics has been the most extensively explored realm of value in recent years; my treatment of it here will therefore be brief. As I indicated, ecologists have engaged in this economic discussion with vigor, and have amassed what may be judged as a cogent case. The dollar numbers (where available) often look persuasive, and, as we know, arguments based on numerical foundations are paramount in Western thinking and discourse. Their force is immediately obvious and, when the numbers are right, compelling. There are, however, traps for defenders of biodiversity who are drawn into this line of discourse.

Of the three aspects of direct tangible benefit we derive from the diversity of life around us—food, materials, and medicines— food is the most obvious and, of course, has the longest history. Until a mere ten thousand years ago, humans across the globe lived in small bands, subsisting as hunter-gatherers over a huge range of environments, just as their ancestors had for at least 100,000 years. (The evolutionary roots of such a mode of existence go back at least two million years.) From the frozen steppes to tropical rainforests, from temperate meadows to baking savannah, from coastal shores to high plateaus, humans exploited a myriad of natural food resources. Diets were usually varied, reflecting a knowledgeable exploitation of all that nature offered in its rich diversity. Ten thousand years ago, people began to develop food production, or agriculture, based on their experience of what could be grown or corralled. Artificial breeding—at first accidental and later deliberate—enhanced the food value of certain plants and animals. As a result, fewer and fewer species accounted for ever larger proportions of people's diets. These days some twenty species of plant provide 90 percent of (vegetable) food in people's stomachs worldwide; and just three—maize, rice, and wheat—amount to more than half of harvested crops.

This astonishing productivity from so modest a collection of plant species may be seen as a triumph of modern agriculture, and indeed has frequently been hailed as such. But this notion rests on thin ice, because the concentration of food production of just a few species—essentially monoculture—makes agriculture vulnerable to massive devastation through disease. A pathogen made virulent through mutation to which current hybrids have no defense could wreak havoc from which multicrop agriculture is protected. This is a simple argument of logic that even economists can understand, and they are beginning to do so. In recent years efforts to incorporate previously unfavored varieties of crop plant have gained momentum.

The seven thousand or so species of plant that have been raised as crops during recorded history offer a pool for such incorporation. But there are at least thirty-five thousand, and probably twice that number, of edible plants available for exploitation. It is folly to ignore such riches, because they may offer varieties that are able to thrive under adverse environmental conditions and in the presence of pathogens that are fatal to conventional crops. For instance, wild varieties of maize and tomatoes that were cross-bred with cultivated species have transformed the agricultural industry in recent years. The stories of the discoveries of these species are now legendary among biologists and agriculturalists. In both cases, the wild varieties existed in tiny populations that might easily have gone extinct in the face of pure chance or economic development and habitat destruction, or remained unknown through ignorance. The technical literature is replete with similar, if somewhat less dramatic, examples. Quite simply, the current level of biodiversity should be valued as a resource from which a more varied and sustainable future agriculture may be assembled.

I'm speaking here of the exploitation of wild plants as crops in their own right, for which there is enormous potential, and as breeding partners through which to confer beneficial characteristics, such as disease resistance, on existing crops. The recently

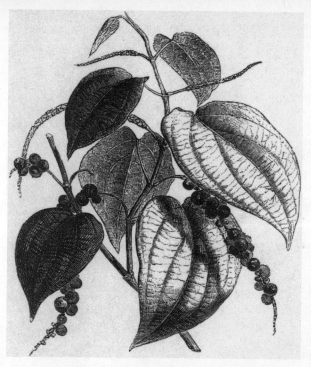

Black pepper plant—Piper nigrum *(Linnaeus).*

honed tools of molecular biology have greatly enhanced the agricultural potential of the genetic resources of wild plants. A decade ago the Cornell University botanist Thomas Eisner offered a graphic analogy of this new potential. A species, he said, is like a book, in which the pages represent its genes, each a repository of genetic information that is the product of millions of years of evolution. It's not a hard-bound volume, however, in which all the pages come as an intact package. It is a looseleaf book, "whose individual pages, the genes, might be available for selective transfer and modification of other species." No longer are agriculturalists limited to conferring the beneficial effects of the genes from one species to closely related species. Through genetic engineering, the myriad genetic libraries that have coevolved since the Cambrian explosion are in principle available for unlimited recombination among unrelated species. Each

The tapioca plant—Manihot utilissima *(Pohl). Technologically primitive societies make use of a vast range of plants, for food and medicine. These days just a handful of plants is exploited, but nature is bountiful with possibilities that are threatened as the world's forests are cut.*

time a species goes extinct, that potential is depleted, never to be regained by our children's children.

It is impossible to put a solid figure on how much of current agricultural output comes from the exploitation of once disfavored species, but it is certainly in excess of half a billion dollars a year. And who can predict the economic impact on mainstream food production of exploiting, for instance, the buriti palm, the maca, the tree tomato, the winged bean, the green iguana, the vicuña, and many, many other wild species of plants and animals, as food sources in their own right and as a rich pool of versatile genes? Some of these species could be raised in large numbers, as is typical of mainstream agriculture, but many are best taken in small quantities from their natural habitats. The notion of limited exploitation of exotic species from their natural habitats is vigorously debated among ecologists. On one hand, there are those who suggest that ecosystems are most likely to be preserved if people can make a living through a limited exploitation of the ecosystem's resources. This is the "use it or lose it" philosophy, which I strongly favor. Others assert that any human impact on ecosystems diminishes them, and their preservation depends on maintaining pristine conditions (whatever that might mean), by preventing all contact or exploitation.

As part of the former philosophy, Charles Peters, Alwyn Gentry, and Robert Mendelson recently made a study of the economic potential of one hectare (two and a half acres) of tropical forest in Peru. The value of such a habitat is usually measured in terms of the timber that can be extracted from it, as a one-time event. But timber is not the only material of value in the forest. In addition to many food resources, there are oils, latex, and fibers (and, less immediately available, medicines). That single Peruvian hectare that Peters and his colleagues studied supports 842 trees, representing some 275 species. (A comparable area of temperate forest may boast half a dozen species at most.) It is from this diversity of species that potentially rich and varied exploitation flows, as follows.

Peters and his colleagues calculated the annual value of the diversity of foods, fibers, oils, and other marketable materials from the forest to be in the region of $400. But this is a sustainable yield, repeatable year by year, so the net present value (an economist's measure) becomes more than $6000. Add to this a selective cutting of trees, and the value reaches $6820. This compares with about $1000 for the clear-cutting of the timber (the usual practice) for the hectare and about $3000 for its use as cattle pasture over a period of many years. "Without question, the sustainable exploitation of non-wood forest resources represents the most immediate and profitable method for integrating use and conservation of Amazonian forests," Peters and his colleagues wrote in a report of their work, in *Nature,* in 1989.[3] An issue of some importance here is that when a hectare of forest is cut and the timber sent to pulp and sawmills, the $1000 return is instant. Two and a half years are required for the same return from a sustainable use of forest products. This may not seem a big difference, but short-term gain, rather than long-term value, is often the measure that is used.

This calculation of the value of a sustainable exploitation of the forest has been criticized variously as being too high and too low, which perhaps means it's about right. Whatever the case, the exercise demonstrates that measuring the value of rainforests solely in currency of timber at the sawmill is shortsighted at best and criminally destructive at worst. The point here, again, is that the value of the forest, if properly exploited, flows from the diversity of species that make up the ecological community. Arguments by economists, like those by Julian Simon of the University of Maryland, that the forest land becomes more valuable if it is clear-cut and replaced with plantations of, say, non-native eucalyptus and pine, are demonstrably wrong.

Of the forms of potential economic benefit encompassed in existing biodiversity, and particularly in the diversity of rainforest species, that of medicines has recently received most play, with good reason. Anthropologists have long known that indigenous peoples use a very large proportion of plant species in their environments, not only as food resources but also in healing practices. I've seen this in my own country, particularly among the Maasai, and have occasionally used the antibacterial, anti-inflammatory powers of the juices squeezed from sansivaria, a succulent plant that grows in arid areas of Kenya. For a long time viewed as quaint and mystical, the use of herbal remedies in traditional medicine has recently come to be seen as a guide to developing powerful pharmaceuticals in Western medicine. In fact, plant products are the source of a significant proportion (some 25 percent) of pharmaceuticals currently used in Western medicine, not the least of which is aspirin. Derived from a constituent of meadowsweet, this modest chemical is the most used medicine in the world. A further 13 percent of Western pharmaceuticals are made from the products of microorganisms, and 3 percent from animals, bringing the total of organism-derived pharmaceuticals in high-tech medicine to close to half.

Many of these remedies, such as penicillin, have been around for decades, and we tend to take them for granted, a mere historical happenstance. Others, such as Vincristine and Vinblastine, alkaloids from the rosy periwinkle from Madagascar, are recent stars. Discovered serendipitously, these chemicals cure patients with the deadly cancers of acute lymphocytic leukemia and Hodgkin's disease. The saga of the rosy periwinkle and its benefits to Western medicine is an example of the good and the bad in the exploitation of natural resources as a source of powerful pharmaceuticals.

Plants are versatile chemical factories, producing a vast range of molecules used in many different ways, sometimes as part of a species' daily metabolism, sometimes as a defense against predators. Plant alkaloids, for instance, are effective deterrents against would-be grazers and browsers. With some 250,000 species of plants in today's world, products of hundreds of millions of years of evolution under a broad range of environmental conditions, the arsenal of alkaloids and other available chemicals is essentially infinite. Vincristine and Vinblastine are just two of sixty such alkaloids produced by the diminutive rosy periwinkle, and they save thousands of lives and net close to $200 million a year in sales. The lives saved represent the value to be enjoyed by exploiting what nature has to offer. This is the good of the rosy periwinkle story. The bad is that Madagascar, the country to which the plant is a native, has received not a penny of the profits that pharmaceutical companies reap from exploiting the genetic heritage of the country.

Madagascar, an exotic island in the Indian Ocean, boasts extraordinarily rich fauna and flora, much of them unique. The country's people might be more willing to slow down or even halt the massive destruction of its forests if they could see benefits in maintaining them as, for instance, a potential source of revenue from medicines discovered in them. The genetic heritage of a country is just as much a national resource as are its mineral ores. Western corporations that exploit that heritage for monetary profits must pay appropriately for them. I'm glad to see that, as Western pharmaceutical companies have stepped up their efforts to find potent medicines in the diversity of plants in tropical regions in recent years, more equitable relationships are being developed. When the pharmaceutical giant Merck agreed in 1992 to pay $1 million to Costa Rica over a period of two years for the privilege of seeking potential drugs in the country's diverse forests, it was a good first step. Others are following this example.

A decade ago, the U.S. pharmaceutical industry spent more than $4 billion a year on research and development of synthetic drugs. At the same time, the sale of prescription drugs deriving directly from natural plant products netted twice that figure, some $8 billion. And yet no company had an active program to find new drugs from higher plants. Now that Western science has come to see plants as smart chemical synthesizers and traditional medicine as more than whimsy, this has changed. With many major companies, and several national research organizations, having joined the search, the list of new, natural medicines can be expected to grow. It takes many years for a chemical to go from demonstration of pharmacological effect to use as a prescription medicine, of course, but the dollar value of such medicines is likely to be enormous as the years pass. And this is the dollar value that may be applied to the ecological communities around the world, which collectively constitute global biodiversity.

As a source of new or improved crops, of materials, and new medicines, the world's biota unquestionably has vast value, something to which dollar numbers can be attached and used in debates with economists. Some ecologists have resisted the recent move to place monetary value on natural habitats as a response to the challenges of Julian Simon and other economists. Hugh Iltis, the University of Wisconsin botanist who discovered a tomato variety that has netted more than $80 million in the last decade, expressed this point strongly at a major conference on biodiversity. "I have no patience with the phony requests of developers, economists, and humanitarians who want us biologists to 'prove' with hard evidence, right here and now, the 'value' of biodiversity and the 'harm' of tropical deforestation," he said. "Rather, it should be for them, the sponsors of reckless destruction, to prove to the world that a plant or animal species, or an exotic ecosystem, is *not* useful and *not* ecologically significant before being permitted by society to destroy it."[4] Many ecologists privately agree with Iltis, but because they feel compelled to meet economists on their own terms, they publicly argue in favor of putting dollar figures on ecosystems as a way of defending the value of biodiversity. It is a tactic fraught with danger.

Although the dollar figures may look powerful, they can never be secure and can never represent a complete defense of the value of biodiversity. Who could seriously argue that the future material benefits I've just described depend on the continued existence of all 250,000 species of plant? We don't know in which of these species lies the cure for AIDS or the genes for important new crops, of course. But even half this number would still represent a vast genetic library from which we may be able to draw any benefit we seek, given the appropriate effort. And what of the species that offer no potential material benefit? Do they have no value? In the economic arena, they don't. As David Ehrenfeld put it, "If I were one of the many exploiters and destroyers of biological diversity, I would like nothing better than for my opponents, the conservationists, to be bogged down over the issue of valuing."[5]

Value in conventional terms changes as circumstances in the world change. Feather quills once had important economic value, because they were ubiquitous writing implements. No longer. The potential of the world's plants as a source of new medicines could vanish overnight with the development of new ways to design drugs. Indeed, with the advent of theoretical chemistry in powerful computer simulations, and the ability to "evolve" chemicals under controlled conditions, the pharmaceutical industry may be already on the verge of a revolution that would leave the tropical forests valueless in its eyes. The ecologist would then be devoid of a major argument for the importance of the forests, and the economist wins. "It does not occur to us that by assigning value to diversity we merely legitimize the process that is wiping it out, the process that says, 'The first thing that matters in any important decision is the tangible magnitude of dollar costs and benefits,' " observes Ehrenfeld. "If we persist in this crusade to determine value where value ought to be evident, we will be left with nothing but our folly when the dust settles."[6]

I agree with this logic and sentiment, but I wouldn't suggest that economics plays no part. For anyone whose daily responsibility is conservation in the real world, as it was for me, economics cannot be escaped. An ecosystem, as a park or preserve, has value in tourism, for instance, and the park must bring material benefits to people living in or near it; otherwise they have no motivation for preserving it. But attempts to place dollar values on ecosystems in a standard economic framework is folly and doomed to failure.

Suppose Julian Simon and like thinkers are correct in suggesting that replacing rainforests with tree plantations or pastures would increase the economic value of the land. Suppose the pharmaceutical industry shortly comes to have no use for the diverse chemical products of hundreds of million years of evolution. Are ecologists left with no argument with which to defend biodiversity? No. Although our dependence on natural products for food and materials is direct and obvious, there is a less direct but no less important benefit we derive from the world of nature around us. The Stanford University biologists Anne and Paul Ehrlich term it "ecosystem services." So fundamental is our reliance on the world's plant communities for maintaining a life-sustaining environment, it is almost embarrassing to have to raise it as an argument about the value of biodiversity. But when Julian Simon suggests that the removal of the world's tropical forests would do us no harm, does he really think that the destruction of more than half the world's species—not just plants, but animals too— would have no effect on the functioning of the world's biota? It seems so.

For as much as a billion years, the Earth's atmosphere has been maintained with high levels of oxygen and carbon dioxide, at first as a result solely of photosynthetic organisms in the oceans and later of similarly functioning organisms on land. Moisture circulates through the same terrestrial agency; for instance, in its lifetime, a single rainforest tree pumps almost three million gallons of moisture from the soil to the air. Rainfall patterns in the rest of the globe depend on this continued process in myriad trees around the equator. The forests are the planet's lungs. Trees—whether tropical or temperate—don't live in isolation, however. Researchers in Denmark recently counted some forty-six thousand small earthworms and their relatives, twelve million roundworms, and forty-six thousand insects under just one square meter of forest floor in their country. One gram of that same soil contained more than a million bacteria of one type, 100,000 yeast cells, and 50,000 fragments of fungus.

These numbers cheat ready comprehension, but the numbers themselves are not what is important. We see in them not just a riotous profusion of different life forms, but a rich pattern of interaction, a living network that *is* the ecosystem. Toward the end of *The Origin of Species* Charles Darwin conjured up a graphic image of this interconnection as a product of evolution: "It is interesting to contemplate an entangled bank, clothed with many plants of many kinds, with birds singing on the bushes, with various insects flitting about, and with worms crawling through the damp earth, and to reflect that these elaborately constructed forms, so different from each other, and dependent upon each other in so complex a manner, have all been produced by the laws acting around us."[7]

An example of this interaction came as a significant surprise when, not very long ago, biologists realized that the ubiquitous subsoil fungi were essential for the daily survival of higher plants. Countless fungal filaments are in close symbiosis with plant roots, making essential minerals available, without which the plants would perish. In every local ecosystem around the world, microorganisms, higher plants, invertebrates, and vertebrates coexist with labyrinthine interdependence, partners in creating and sustaining the physical environment of atmospheric gases and soil composition and chemicals. Individual ecosystems work as integrated wholes, not as species in the company of but isolated from other species. *Homo sapiens* is one unit in that pattern of interdependence.

Two decades ago, the British chemist and inventor James Lovelock took the notion of ecosystem interdependence a step further; he took it to the global level. Termed the Gaia hypothesis, his suggestion was that all the ecosystems of the planet were essentially interdependent, operating as a whole and inextricably linked to the physical environment. A consequence of that interdependence was the establishment and maintenance of the physical conditions necessary for life. Some Gaia enthusiasts extrapolated the theory to the point of suggesting that the Earth's biota is like a single organism, one with the purpose of maintaining itself. For this reason, the hypothesis was not taken seriously by biologists; it all seemed too mystical. Recently, however, ecolo-

It was only recently discovered that plant roots are enmeshed in a vast network of fungal filaments, upon which they depend for sustenance. This is an example of interaction among organisms, which is a central aspect of past and present diversity.

gists have cut through the mystical fringe and acknowledge that Lovelock was correct: just as the viability of individual ecosystems is maintained through the interaction of the species that compose them, so the environmental health of the Earth's biota flows from an interaction of all its ecosystems.

The importance of Gaia theory for our appreciation of biodiversity is profound. The theory's author told a major conference on the subject: "No longer do we have to justify the existence of humid tropical forests on the feeble grounds that they might carry plants with drugs that could cure human disease. Gaia theory forces us to see that they offer much more than this. Through their capacity to evapotranspire vast volumes of water vapor, they serve to keep the planet cool by wearing a sunshade of white reflecting cloud. Their replacement by cropland could precipitate a disaster that is global in scale."[8] Predictions of global disaster aside, the core of Gaia theory has now been tested many times, and vindicated. The balance of many chemical cycles, not just moisture, has been shown to flow directly from the functioning of ecosystems. Although there are many who still talk about the theory in mystical terms, Gaia has become serious science, and we are forced to take note of its implications.

Beyond recognizing that the maintenance of physical conditions that sustain life as we know it depends on the Earth's biota working as an integrated whole, what can we say about biodiversity? To reiterate a question I asked earlier, but in a different context: Do we need all fifty million extant species for Gaia to remain operative? Would 100,000 species of plant do the job, rather than the quarter of a million that exist now? Are all 600,000 species of beetle really necessary?

The world's ecosystems will operate as a whole so long as each is able to persist, especially in the face of such occasional perturbations as storms and fire. If that persistence—or stability—has to rest on a foundation of high diversity of species, then biodiversity as we currently experience it can be judged to have value. If, however, stability does not require a rich species diversity, then we would be unable to value current biodiversity in this way. (There may, of course, be ways other than ecosystem stability in which high biodiversity is essential.) The diversity-stability equation has been debated by ecologists for years, without agreement, particularly between field researchers and theoreticians. Resolution may, however, be close.

Field biologists have traditionally believed that the complexity of interaction among species within ecosystems is important for their stability. That belief was based more on intuition than on demonstrated fact, for two reasons. First, doing ecological experiments in nature is notoriously difficult; the scale of things is usually defeating, both physically and temporally. How do you alter a natural ecosystem to order, set up controls, and then wait half a century for results? So, there were few real observations on which to base judgements. Second, as a field biologist, once you are immersed in the ecosystem you study, there is a very real sense that everything has a part to play in the emergence of the whole. This "in the bones" feeling influenced biological thinking for a long time, but in the early 1970s, theoretical models appeared to imply that the fewer species an ecosystem contained, the more stable it would be. The models, developed principally by Robert May, of Oxford, essentially said that the more components there were in the system, the more things could go wrong. And if the components were tightly interconnected, something going wrong in one part of the system could initiate collapse of the whole. Intuition versus powerful mathematical models; the stand-off continued for years.

As the debate continued, biologists kept asking: If ecosystems don't need a rich diversity, why are they so rich in species? Two views emerged, the rivet-popper hypothesis and the redundancy hypothesis. Developed by the Ehrlichs, the former says that each species plays a small but significant part in the working of the ecosystem, like each of the many rivets that hold an airplane together. The loss of a few species—like the loss of a few rivets—weakens the whole, but not necessarily dangerously. Lose some more, and catastrophe looms, especially if the system faces a severe test, like environmental perturbation for the ecosystem or air turbulence for the plane. In the redundancy hypothesis, proposed by the Australian ecologist Brian Walker, most species are seen as superfluous—more like passengers on a plane than rivets holding it together. Then only a few key species are required for a healthy ecosystem. Several of the many species in the system can play the key roles. Which model is correct?

James Lovelock stepped into the arena with a theoretical model of his own, in the form of a computer simulation—Daisyworld—that built miniecosystems. According to the output from Daisyworld, the more species there were in the ecosystem, the greater its stability. This seemed to support a direct relationship between biodiversity and stability, the rivet-popper hypothesis. Theoreticians were reluctant to accept Lovelock's conclusions, but had to listen when results from innovative ecological experiments became known in 1993 and 1994. Researchers in England and the United States independently tested the effect of diversity on the productivity and stability of ecosystems.

Productivity is simply the quantity of living material an ecosystem can generate in a given period of time. This is just as important for agricultural systems as it is for natural ecosystems. Michael Swift, a biologist at the United Nations Tropical Soil Biology and Fertility Program in Kenya, has demonstrated convincingly the benefit of species diversity in agricultural systems. The best way to increase productivity in a maize field is by adding melons, trees, and nitrogen-fixing beans, not by squeezing in more maize. In their experiments at Imperial College's field station in England, John Lawton and his colleagues also found that productivity is boosted by species diversity. The result makes immediate sense, once you see it. Individuals of a single species will compete for the same resources, particularly space. Individuals of different species—some small in height, some medium, some tall —can take advantage of different spatial territory. More of the available space is used, so more individuals are supported, giving higher productivity.

Does this explain the high species diversity in most natural ecosystems? Only up to a point, beyond which productivity seems to level off. For instance, although the diversity of tree species in East Asian forests is about six times that in North American and eight times that in Europe, productivity of all the forests is similar. Productivity is therefore only part of the answer. Stability may be the rest.

In what is surely one of the most important discoveries in a long time, David Tilman, of the University of Minnesota, and John Downing, of the University of Montreal, found a direct link between species diversity and the health of a natural ecosystem. They conducted an eleven-year study of native grassland ecosystems in Minnesota, which, as luck would have it, included the worst drought experienced in the area for fifty years. What might have been a catastrophe for the experiment turned out to be a blessing, because it highlighted clearly the difference between ecosystems rich in species and those which were poor. The former suffered significantly less loss of species and productivity in the face of the drought, and recovered much more rapidly. "Our results . . . support the diversity-stability hypothesis, and show that ecosystem functioning is sensitive to biodiversity," they stated in *Nature* in January 1994. "Our results do not support the species-redundancy hypothesis, because we always found a significant effect of biodiversity on drought resistance and recovery."[9]

Species do appear to be more like rivets in a plane's structure than passengers in its seats. But, again, exactly how many rivets can be lost before the plane becomes endangered is unknown, and no one has a way of finding out. Nevertheless, biologists recently made a strong statement in support of the diversity-stability relationship, which effectively raises a flag in support of high species diversity. The Scientific Committee on Problems of the Environment, a part of the United Nations Environment Program, met in California early in 1994 to review evidence and opinions on the matter. Displaying the courage to admit that no one had a complete answer, the committee concluded that high species diversity *is* beneficial; there may be some redundancy in many ecosystems; and who could claim to be sufficiently without ignorance to have the confidence to tamper with it?

I've always been passionate about wild and remote places and have had an intense interest in and love of animals. As a teenager I wanted nothing more than to be a game warden, out in the wild, trapping dangerous animals, leading a life of adventure. When I became director of the wildlife service, I not only tackled the practical issues of the conservation of wildlife in the face of development and the daily horrors of seeing the spoils of poaching elephant, but I was also brought close to a deep, visceral passion for nature.

By accompanying my parents on their searches for early human relics, as a child I absorbed a deep knowledge of animals and their natural environment. I also learned how to fend for myself in the wild; how to find water and food in what looked like a barren desert; how to track and trap wild animals. Unconsciously I learned how to be part of nature, to respect it, not be afraid of it. Although I didn't realize it at the time, I was extremely fortunate in my childhood experience, because it allowed me to connect with something that is fundamental to the human psyche. Most people are denied this, although, unconsciously, most strive for it.

For some 150,000 years, our *Homo sapiens* ancestors lived as hunter-gatherers, in many different environments. This highly successful mode of existence had its origins with the evolution of the genus *Homo,* sometime prior to two million years ago. The expansion of the brain that has occurred since that time, and the development of the human psyche that has gone along with it, were in the context of the hunter-gatherer way of life. It was a life of extreme intimacy and dependency upon all of nature. It required keen sensitivity to every aspect of nature. Our ancestors undoubtedly saw the other species in their world as a source of food, of many kinds; they must have witnessed much to wonder at in that world, as we see reflected in the cave and rock shelter paintings of Europe and Africa; and they knew themselves as an integral part of this diverse world. I have written often that, although we occupy a modern technological world, we have the minds of hunter-gatherers. I knew this instinctively, if not explicitly, when, as a boy, I listened, fascinated, to my father's stories; and when I wandered confidently in the wild terrain of Olduvai Gorge. A decade and a half ago, Edward Wilson put a name to this instinct: he called it biophilia.

Recently, Wilson defined biophilia as "the innately emotional affiliation of human beings to other living things."[10] Wilson is speaking of something deep within the human psyche, something that has become a part of our very existence through millions of years of evolution. He is speaking of emotional responses that touch the essence of humanity, the very essence of our history. Some of these emotional responses to nature may be negative, as is the aversion many people have to snakes, even to the *idea* of snakes in the abstract. But many are positive. Why else do people so often seek relief from urban stress by visiting wilderness areas? Why else do people in the United States and Canada more often attend zoos than all major sports combined? And why do people secure a home in the countryside if they have the means to do so? The psychologist Roger Ulrich has shown that, given a visual choice between urban or rural scenes, people overwhelmingly prefer the latter.

The preference may go even deeper, however, perhaps reaching back to our ancestral past. Given visual choices among rural scenes, people show an overwhelming preference for rolling landscape vegetated with scattered trees, preferably flat-topped trees. The University of Washington ecologist Gordon Orians interprets this preference as a deep psychic connection to our origins in East Africa, the preferred landscape being reminiscent of woodland savannah. Some find this suggestion far-fetched, but, like people's spontaneous love of some animals and fear of others, it strikes a chord that cannot be ignored. Whether negative or positive, our response to wild nature, according to the biophilia hypothesis, is an ineradicable part of human nature. It is the heritage of eons spent as hunter-gatherers in ancestral times.

Western culture, with its high-tech civilization, has come to ignore the essential connection between the human psyche and the world of nature, while emphasizing the promise of worlds beyond our own planet or solar system. It ignores the connection, but the connection is still there. Other cultures do not do this. Half a century ago the Native American Luther Standing Bear wrote: "We are of the soil and the soil is of us. We love birds and beasts that grew with us on this soil. They drank the same water and breathed the same air. We are all one in nature. Believing so, there was in our hearts a great peace and willing kindness for all living, growing things."[11] Western culture has come to view *Homo sapiens* as not only special in the world (which we undoubtedly are in many ways), but also separate from that world. It is as if we were set down on the Earth, complete and finished in our present form, to have dominion over Earth's creatures. This is not true, of course, but it is all too easy to think in evolutionary terms, seeing *Homo sapiens* as the product of a long process, and yet *still* perceive us as special and separate. There is, after all, nothing like us in the rest of Creation. There is, after all, a tremendous gulf between the mind of *Homo sapiens* and that of our closest relatives, the African apes.

But if one spends one's life reconstructing the path taken by our distant ancestors on their evolutionary journey from ape to human, the gulf disappears. Not only is it possible to find and identify *physically* the species that connect us to our ancestral roots—*Homo erectus* and *Homo habilis*—but we can also build a picture of their *behavior*. Most important of all, we can see the context in which our evolution took place, the constantly shifting ecosystems of which our ancestors were an integral part. It is this intimacy that impressed itself on the emerging human psyche. It is this intimacy to which Luther Standing Bear instinctively alluded. It is this intimacy each of us experiences today, in different, perhaps muted ways. And it is this intimacy that enables us to place value on the biodiversity of which we are a part today, separate from the direct economic benefits of foods, materials, and medicines, and separate from the ecosystem services upon which our physical survival depends. The value of the species around us now reaches to the human spirit—not an easy thing to say in the context of science, but valid nonetheless.

We may value biodiversity because it nurtures the human psyche, the human spirit, the human soul.

The Balance of Nature?

ECOLOGICAL COMMUNITIES do not exist in a benign harmony, but, instead, are shaped by many forces, some of them chaotic, some random. Above all, there is constant, dynamic change.

Humans have had a major impact on such communities in the historical past, revealing how vulnerable such ecosystems are to disruption by an invading species.

The plight of the modern elephant reveals not only our potential impact on nature, but also the challenges that we face in trying to protect complex ecological systems.

9

Stability and Chaos
in Ecology

I F YOU WERE TO TRAVERSE the globe from pole to equator, you would see what has been called "nature's infinite
variety" writ large. From frigid tundra and Alpine meadows,
through temperate woodland and pastures, to tropical forests
and open savannahs, you would experience a huge range of different ecological communities. This difference is an important
motivation for North Americans and Europeans to visit my own
country, Kenya, each year. The contrast between the plant, animal, insect, and bird communities in the home and visited environments is enormous. Not only does the number of species in
ecological communities increase as one moves from high to low
latitudes, but the types of species in those communities change
too. (Polar bears are not to be found as members of tropical
ecosystems, nor are large primates—humans aside—seen as natural components of temperate or Arctic fauna.) This large pattern
is, of course, partly a consequence of the adaptation of species to
local conditions, the most noticeable of which are temperature
and humidity. A local ecological community, therefore, is a col-

lection of species that in the midst of their individual differences have a common adaptation; that is, to local environmental conditions.

You don't have to be a globe trotter to see variety—great, if not infinite—in nature, however. Anywhere you may find yourself in the world you would be among ecological communities that differ from one another, sometimes profoundly so. I've already described the dramatically contrasting ecosystems that cloak the Great Rift Valley, in Kenya. Again, it is easy to recognize an important source of this diversity here: from the floor of the valley to the peaks of the rift's high walls, a myriad of microclimates prevail, offering contrasting conditions for different kinds of species. Evolution and adaptation work at all scales, producing patterns at all scales.

One of the goals of ecology, as the University of New Mexico ecologist John Wiens has put it, "is to detect the *patterns* of natural ecosystems and to explain the causal *processes* that underlie them." In what I've described so far, the major process is, I've suggested, species' adaption to local conditions. But anyone who has strolled through a wood or across a meadow, and is observant, knows that he or she is passing through a patchwork of nature, not through complete uniformity. One kind of tree will be common for a while on the walk, only to be absent later; one type of flower will be rare in the south end of the meadow and abundant in the north. This patchwork of nature is a patchwork of similar though distinct ecological communities. What are the processes that have shaped *these* patterns? There's nothing obviously different about the environmental conditions at the two ends of the meadow, so why are the ecosystems there not identical? Perhaps even the keenest observer might miss crucial environmental differences, however, such as soil chemistry or variations in the level of the water table. This kind of explanation is popular among some ecologists, as explicated recently by Seth Rice, a biologist at the University of North Carolina. "All environments in all ecosystems are variable in space and time," he wrote in a major review article. "The environmental patchiness is based on physical and chemical gradients that are ubiquitous."[1] In other words, the patterns in the biological realm are determined by, and reflect, the underlying patterns in the physical world, through the process of local adaptation.

Or are they? Although it seems intuitively reasonable, even obvious, that ecosystems should be molded in this way, in recent years it has become apparent that other forces are at work, too, ones that are much less obvious—even intuitively *un*reasonable.

The territory into which we are straying here—that of community ecology—poses what are probably the science's most important and least tractable problems. At its core there is a single, simple question: How does an ecological community come to be the way it is? One response, long popular, is that the community is the way it is because it must be so, because local conditions dictate it to be so. More important, the members of the community are considered to be tightly adapted to these conditions in concert and interact so intimately with one another—are in reality dependent upon each other—that a community of different species' composition could not survive. This is stating the case perhaps a little strongly, but it catches the essence of much of recent thinking in ecological theory. Associated with this view of the fundamental structure of ecological communities is the notion of the "balance of nature."

Again, there is something intuitively reasonable about the phrase and its implications, something even reassuring. If ecosystems are the way they should be, then it follows that, when disturbed by some means, nature will work quickly to restore them. Half a decade ago, Fairfield Osborn, son of Henry Fairfield Osborn, encapsulated this sentiment in his book, *Our Plundered Planet:* "Nature may be a thing of beauty and is indeed a symphony, but above and below and within its own immutable essences, its distances, its apparent quietness and changelessness it is an active, purposeful, coordinated machine." The machine functions to maintain communities in their balanced state. The phrase "balance of nature" became a powerful metaphor in ecological and lay circles for a perceived fundamental natural harmony, evoking the rightness of the world as we experience it. As a result, ecology sported a distinct air of mysticism for a while. Even when that mysticism disappeared, as it did a couple of decades ago, the phrase remained. Stripped of any notion of purposefulness, the balance of nature came to refer to an ecological community's ability to resist or recover from perturbation, which was given the more objective term *stability*. Whether one uses "balance of nature" or "stability," there is, Stuart Pimm has noted, "something unmistakably fuzzy about the terms."[2]

How an ecological community assembles itself remains a major question, as do the behavior and characteristics of the community once it is formed. These issues are immensely complex, because they involve many variables (that is, individual species) that may interact in many different ways—and all of this is set in an often tumultuous physical environment. As a result, the range of potential patterns is immense, even infinite, and trying to understand why certain patterns emerge rather than others is a daunting challenge.

This may sound all very academic, and in a sense there *is* a strong urge to understand how nature works and what was the source of the biodiversity of which we are a part. But that understanding is also vital to our desire to protect biodiversity, to conserve nature's infinite variety. For instance, in addition to being able to account for the different species' compositions of communities, we need to understand what makes species' populations fluctuate within communities. We need to understand what makes one community vulnerable to perturbation, particularly human-imposed perturbation, while another is resistant. We need to know why some communities recover quickly from disruption while others do so slowly. We need to know why alien species can invade some communities easily, but not others, and the consequences of such invasions. We need to understand which species are vulnerable to extinction and which play such vital roles in their community that their extinction provokes a cascade of further extinctions. Each of these questions finds a place of equal importance in an academic ecologist's text and a conservationist's handbook.

This chapter begins with an exploration of the recently discovered, and counterintuitive, insight into the reasons that species' populations fluctuate the way they do; that is, sometimes regularly and sometimes wildly and erratically. This brings us face to face with the realization that much of the time nature is not in balance at all, but instead is chaotic. For some, this may be too disconcerting an image of nature to accept, as it seems stripped of any fundamental harmony. Then I will discuss the ways ecologists have come to understand something of the processes by which communities form; for the most part, researchers are forced to use powerful computer-generated models of ecosystems to do this. Here we will see that ecological communities *seem* to have a mind of their own, because they improve themselves over time, becoming ever more resistant to invasion by alien species. I will talk about the dynamics of species invasions—about what makes a successful invader and what determines the impact of such events. This is an important issue for conservation. Lastly, I'll give a cautionary tale about how conservationists should maintain the stability of ecosystems—that is, by allowing them to change. Again, this is counterintuitive, as is much that is being learned about community ecology. Nature is not all she seems.

In the Smithsonian Institution's National Museum of Natural History, in Washington, there's an exhibit that never fails to make onlookers' flesh creep: every surface of a mock-up kitchen swarms with cockroaches, hundreds of thousands, perhaps millions, of them. Those spectators who manage to go beyond the immediate visceral revulsion and read the exhibit's caption learn that the entomological horde before them is the descendent family of a single female throughout her lifetime—in theory, at least. Luckily, such potential fecundity is rarely realized. As Darwin observed in *The Origin of Species,* most organisms have the potential to leave more offspring than actually survive. Something keeps this potential in check. (Those not fond of cockroaches should be grateful.) That "something" includes limited food resources, competitors, predators, climatic insults, disease, and other agencies. Although over a long period of time the average number of individuals in a population may be relatively steady, in the short term these numbers bounce around the average. Some of these fluctuations are gentle, some are dramatic, even wild, with population explosions and crashes, or booms and busts, as ecologists call them.

A key challenge to understanding the dynamics of ecological communities in the short term—that is, one or a few decades—is to gain an insight into precisely what drives the fluctuation of individual species' populations within them. As Robert May has pointed out, "Such an understanding is not only fundamentally important, but it also has practical applications in trying to predict the likely effects of natural or man-made changes such as occur when a population is harvested or when climate patterns alter."[3]

Under the rubric of the balance of nature, population fluctuations are readily explained, in principle at least, if not in detail. Species' populations and the communities of which they are a part are viewed as being at or close to equilibrium. Left unperturbed by external interference, the interactions between the plants, herbivores, and carnivores that constitute a community reach a steady state—equilibrium—with species' populations in careful balance. It is Fairfield Osborn's "coordinated machine" humming smoothly. The limited food resources, the competitive interactions, the toll of predators, and even the effects of disease are all part of the workings of this coordinated machine. Once a community of species reaches equilibrium, the major force for disturbing the balance is climate, either in long-term shifts or sudden, forceful episodes, such as storms and temperature changes. Climatic shifts will favor some species and be detrimental to others. If, for instance, a storm decimates the population of a certain plant species, the herbivores that depend on it as a source of food will also suffer; and in their turn, predators that prey on those species will find their food resources reduced, causing a population collapse. At the same time, other prey species of these predators may have a better chance of survival, leading to a rise in their population numbers. A single storm can therefore cause the populations of some species to boom while those of others bust.

Equilibrium is temporarily lost as different species' populations fluctuate. Eventually, however, after populations swing back and forth across the point of balance, equilibrium will be restored—until another external perturbation intrudes. Because ecological communities are rarely blessed with long periods of freedom from external buffeting, populations fluctuate for much of the time. The classic example in the ecological literature is the recent history of the Canadian lynx. For more than two centuries, between 1735 and 1940, the lynx was trapped for its fur, and the record of pelts recovered by trading companies provided ecologists with an unprecedented set of data for studying the species' population history. A strong pattern is easily discernible in the data, showing that the lynx population went through episodes of dramatic booms and busts. For instance, between 1830 and 1910, the population peaked every nine or ten years, after which it collapsed rapidly. The pattern repeated itself with some regularity, but the population peaks varied considerably, between about ten thousand and sixty thousand individuals. When ecologists first analyzed this history, they assumed that the pattern emerged from a predatory-prey interaction, between the lynx and the snowshoe hare, its principal prey. When a predator has culled the prey population to low levels, its food resource is dramatically diminished, and the predator population falls, too. With predation pressure now reduced, the prey population can recover; the predator population follows in its trail. This regular oscillation is how it looked with the Canadian lynx and the snowshoe hare.

It turned out not to be so simple. The hare population gyrated in size because of fluctuations in *its* food supply, and the lynx population appeared to be tracking it. This scenario has a logic to it, and it made the chain of interaction longer. But even so,

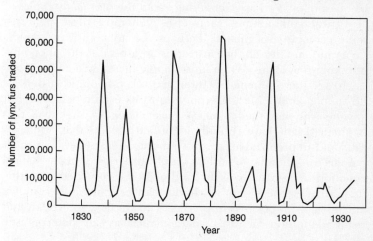

The lynx is the classic example of a species with regularly oscillating numbers, as shown here. It was once believed that lynxes were partners in a dynamically unstable association with their main prey, the snowshoe hare. Recently it has been recognized that the cycle is driven instead by the interaction between hares and their food plants, with the lynxes being carried along more or less passively by changes in the abundance of hares.

the pattern was not perfectly regular, and even looked erratic in parts. In fact, this combination of some regularity and some apparent randomness is typical of population fluctuations of many species. Insect plagues follow this pattern; so, too, do the population eruptions of sea urchins in the northwestern Atlantic, and of Dungeness crabs in the Pacific Northwest. Wherever you look at ecological communities, not only do you see populations fluctuate, but you also see apparent randomness in these fluctuations. From marine plankton to elephants, from moths to mice, the picture is the same. How is this explained from the point of view of the balance of nature and population equilibrium? The answer is simple: everything that is seen in the history of populations—whether it is regularity or apparent randomness or any combination of these—is said to be a direct result of *external* forces, such as changes in climate. The fact that population history may at times be unpredictable reflects the complexity of those perturbations. Or so it was argued.

Beginning about two decades ago, this interpretation began to be questioned. Perhaps something about the *internal* dynamics of ecological communities themselves generates these patterns, some ecologists speculated. Perhaps the apparent randomness isn't random after all, but instead is an aspect of the phenomenon known as chaos. When most people hear a system described as chaotic they immediately infer that it is random, that there is no simple, analyzable underpinning to its erratic behavior. But anyone who has read James Gleick's book *Chaos* knows that mathematicians have recently identified systems that, while erratic and unpredictable, are not random. The behavior of such systems can often be described simply, in terms of mathematical equations. The paradox is that, although it is governed by mathematical rules, the system's behavior can be immensely complex and virtually impossible to predict. This, in fact, is a rough and ready definition of what mathematicians call deterministic chaos.

Such systems have now been identified and analyzed in many physical systems, such as weather patterns and turbulence in fluid flow. Few people realize, however, that population fluctuation in ecological communities was among the first phenomena to be studied as potential sources of chaotic behavior. That was done twenty years ago, by Robert May, and described in a classic

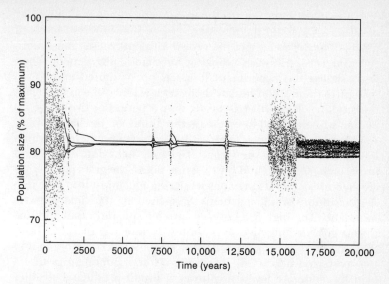

The diagram shows the computer-simulated history of a population of Dungeness crabs. Notice that from time to time the population size fluctuates wildly, even though no external trigger is involved. This is one illustration of the unexpected and unpredictable property of chaos in living systems. (Courtesy of Alan Hastings and Kevin Higgins)

paper in *Nature*. Biologists have been slow to venture down the path that May identified, partly because of the strong adherence to the notion of the balance of nature and populations at equilibrium, and partly because biological systems of this kind are far more complex and difficult to analyze than any physical system. As May once wrote, "To some ecologists [chaos] has an air of black magic."[4] Obsessed as they were with the notion of equilibrium, ecologists continued to look for evidence in its support, while routinely ignoring erratic behavior that implied something else was going on.

In the past year or two, however, long-sought evidence of true chaotic behavior in ecological communities has been discovered, in field experiments and in theoretical models. We are now forced to take a very different view of the world of nature and what shapes the patterns we see and experience. It is deeply counterintuitive, and therefore difficult to accept.

In the mid-1980s, David Tilman, an ecologist at the University of Minnesota, asked the question: How would different levels of nitrogen in the soil affect the growth of pant-creeper, an American wild grass? He wasn't looking for chaotic phenomena when he designed the experiment, but was open-minded enough to recognize them when he saw them. At low levels of soil nitrogen, the amount of growth was steady over a period of five years, no matter whether seeds were scattered thinly or thickly. At high nitrogen levels, however, a very different picture emerged, which incorporated the classic signal of chaotic behavior: erratic and unpredictable boom and bust. At one point, the grass population crashed to $1/6000$ of its previous level, sinking almost into oblivion. A description of what actually happened in the field appears mundane. The high level of soil nitrogen spurred rapid and luxurious growth; this died over winter, producing thick leaf litter that impaired growth the following spring; hence the boom and bust pattern, with varying intensities. More moderate growth, fueled by moderate levels of nitrogen, would produce a steadier population history. When Tilman published these results at the end of 1991, he provoked a mixture of consternation and excitement.

This is an area of ecology where theoretical work is strong, not least because experiments of the nature that Tilman conducted are not easy to design and implement. As I said earlier, experimental manipulation of ecosystems is notoriously difficult. The year Tilman reported his findings, Robert May and two colleagues announced results of a mathematical model (of a parasitoid and its host) that behaved very much like Tilman's system. It showed erratic population swings over a period of many "generations" as a result of interaction between the "species," with no external perturbation. The model described the species and their interaction in mathematical equations. The complex dynamics of the system flowed from *within* it; they were not imposed from *without*. Equally important here is that apparently erratic, unpredictable behavior resulted from mathematically simple relationships; that's the signature of true chaos. More recently, Alan Hastings and Kevin Higgins, of the University of California at Davis, saw something similar in their model of Dungeness crab populations. Again employing simple mathematical equations, the researchers described the species and its behavior along a stretch of theoretical coastline. And again the behavior was erratic and unpredictable, with periods of stability and periods of wild population fluctuations. "Population eruptions may be an underlying feature of the dynamics without any change in physical or biological condition," they noted in their report in *Science* early in 1994.[5]

The demonstration that the size of a population may vary dramatically and unpredictably as a result of interactions within the system, and in the absence of external change, was a major step in understanding the patterns we see in the world of nature. "The concept of chaos is both exhilarating and a bit threatening," noted William Schaffer and Mark Kot, University of Arizona ecologists who have done much to promote the understanding of chaos in ecosystems. "On the one hand, it offers a deterministic alternative to the idea that population fluctuations are solely the consequence of external perturbations. At the same time, chaos could undermine the conceptual framework of contemporary ecology."[6] The concepts of ecology are surely going to be battered by these insights, but from the point of view of biodiversity, chaos is a positive force. As I explained, stable communities can become dominated by one or a few species. In contrast, population fluctuations can drive communities to higher levels of species' diversity. We can see, therefore, that the erratic behavior that flows from the internal dynamics of ecological communities is a force for promoting diversity.

As surprising as this discovery was, chaos apparently holds more surprises for ecologists. I started the chapter by describing the patchwork nature of many environments, with the patches being similar but distinct ecological communities. The conventional explanation for the differences is that they reflect small but important variations in the physical conditions of the environment. Chaos offers another explanation. In their modeling of parasitoid and host pairs, Robert May and his colleagues discovered that not only do populations fluctuate in *size* through time, but their distribution through *space* may be unequal—or patchy —too. Working with models of three species or more distributed over a theoretical, uniform landscape, they found that the dynamics of interaction often kept species separate from each other. "At its most extreme, this . . . [generated] small, relatively static 'islands' within the habitat, giving the appearance of isolated pockets of favorable habitat," they observed in their paper in *Nature,* published in the summer of 1994.[7]

Variation in the distribution of species across habitats (like the trees in the woods and the flowers in the meadows) is common and has been explained by differences in competitive and dispersal abilities, as part of the response to patterns of physical conditions in the habitat. In the counterintuitive perspective of chaos theory, this no longer holds as a complete answer. The patchiness we see in the world can flow from the internal dynamics of the ecosystem, even when the terrain on which the different communities live is exactly the same.

We can now see, therefore, that the world of nature is not in equilibrium; it is not a "coordinated machine" striving for balance. It is a more interesting place than that. There is no denying that adaptation to local physical conditions and such external forces as climatic events helps shape the world we see. But it is also apparent now that much of the pattern we recognize—both in time and space—emerges from nature herself. This is a thrilling insight, even if it means that the work of conservation managers is made more difficult. It was once believed that population numbers could be controlled by managing external conditions (as far as is possible). This must now be recognized as no longer the feasible option it was imagined to be. It is far better to understand and accept the world of nature in its infinite variety and its infinitely complex processes, acknowledging the near futility of attempts to control them, than to imagine through ignorance that it is possible to do so.

In 1789 the Reverend Gilbert White, a clergyman in the south of England, published his wonderful little book, *The Natural History of Selborne.* The fourth most reprinted book in the English language, White's modest tome is a collection of keen observations of nature in and around the village where he lived and ministered. Interest in the behavior of individual species goes back at least to the time of Aristotle, but the Reverend White was the first to focus on species as components of communities. He didn't use the language of modern community ecology, of course; he didn't speak of assembly rules, food webs, and trophic levels. But his recognition of the interactions between species, with their different levels of dependence on one another, is central to modern community ecology's most challenging question: What governs the way an ecological community comes to be the way it is? A crude dichotomy offers itself as a response here: Is it by design or by chance? More specifically, we can ask: Is there something special about the species in a particular community, such that it and it alone is the optimum assembly of species for that habitat? In other words, what kind of order underlies Darwin's "entangled bank"?

There is no easy way to answer this by looking at the real world, because the scale—temporal and spatial—of ecosystems defies easy analysis. For this reason, much of the groundbreaking work in community ecology goes on inside computers, where researchers manipulate experimental ecosystems. Such systems are simple compared with the natural world, but in recent years powerful messages have flowed from them, and, as with the insights from chaos, they are distinctly counterintuitive.

A decade ago, for instance, Stuart Pimm and Mac Post, at the University of Tennessee, set out to assemble one such ecosystem, adding one species at a time (plants, herbivores, and carnivores). Each species was described mathematically, with a suite of behaviors that included its type, its size, and its requirements in terms of territorial range and food resources. Pimm and Post were essentially doing in a computer what occurs in nature when virgin territory is colonized, such as after a fire or on a volcanic island. An ecological community is slowly assembled, in a process called succession, beginning with the simplest of organisms that can thrive in an impoverished habitat and gradually including more species that may depend on those already present. You can't have a herbivore until you have established plant species, for instance; and you can't have predators until prey species are thriving.

In the computer model, species were randomly added to the community. There was no attempt to build a particular community; rather, the idea was to allow a community to develop as it would. And, just as in nature, plant species had to precede herbivores, and herbivores predators. The dynamics of building the community were quite striking. At first, species could be added easily (provided they were ecologically logical). But as the community grew in size (that is, the total number of species), new species found it more difficult to become part of the community.

By the time the ecosystem had about twelve species, invasion was quite difficult and, if successful, often caused the loss of one or more previously established species. This was reminiscent of Darwin's wedge analogy, in which species were viewed as being tightly packed, so that the driving in of a new wedge pushed out an existing wedge. In ecological terms, what Pimm and Post were seeing was the success of alien species invading an existing community, and the effects on the community. Success was easy in species-poor communities and difficult in species-rich communities. The British ecologist Charles Elton had suggested more than three decades ago that this pattern would be seen in nature. Why should this be so?

A traditional explanation is that as the number of species in the community increases, the ecological niches become filled. A potential invader seeking a niche that is already filled will have more difficulty in establishing itself than one whose niche is empty. In the first case, the invader would seem to have to outcompete the established species in order to become part of the community; in the second case, the invader faces no such competition. This seems ecologically logical, but apparently it is wrong. The challenge for the potential invader is not with the established species occupying its preferred niche, but with the community as a whole. This was very clear in a computer model of communities constructed by Ted Case, of the University of California at San Diego. He constructed several different computer communities, and manipulated the degree of interaction among the species in each of them; in some, interaction was strong, while in others it was weak.

"Communities of many strongly interacting species limit the invasion possibilities of most species," he wrote of the work. "These communities, even for a superior invading competitor, set up a sort of 'activation barrier' that repels competitors when they invade at low numbers."[8] If the niche hypothesis was correct, then a potential invader that is competitively superior to an established species would be expected to succeed. And yet it doesn't. Communities with strongly interacting species are less vulnerable to invasion by alien species than those with weakly interacting species, even when the would-be invader is a superior competitor. "These models point to community-level rather than invader properties as the strongest determinant of differences in invader success rates," Case concluded.[9] If correct, this result is extremely important not only for a deeper understanding of the dynamics of ecosystems, but also in the realm of conservation. Frequently, conservation managers are faced with trying to preserve a species in a community that is competitively inferior to an invading exotic species. From Case's models, it is evident that the inferior competitor has the best chance of surviving if the community of which it is a part is species-rich; that is, intact and undisturbed. Preventing disturbance of the community as a whole therefore offers security for its weakest members, by creating a protective network within the community.

The phrase "protective network" has an unmistakably mystical ring to it, so we have to understand from where it derives. The answer is food webs, which have been described as "the road-maps through Darwin's famous entangled bank . . . [which show] how a community is put together and how it works."[10] The maps reveal the interactions between species in the community, such as who eats whom. Biologists have long been fascinated by these maps. Often, the food-web patterns look bewilderingly complex, and initially biologists believed that each community had a unique food web. As they cut through the superficial complexities, however, biologists came to realize that all food webs are very similar, no matter what kind of community they represent, and display the same few common properties. These include the length of food chains—that is, a description of who consumes whom in the community—and the ratio of predator species to prey species. Wherever you look in nature, similar patterns exist. The fact that such commonalities among disparate communities are to be seen even where there is the potential for limitless variety suggests something fundamental about the underlying order in nature. That order seems to emerge from the internal dynamics of the system itself, and is not imposed by external circumstances.

The interactions between species that Ted Case manipulated in his computer model represent food webs of the real world. The protective network that he saw in the communities with strong interactions can therefore be explained as a property of the underlying food-web pattern. No mystical force need be adduced to explain the counterintuitive observation he made.

• • • •

The ecosystems that assembled themselves in Stuart Pimm and Mac Post's computer models displayed networks of interactions between species that closely resemble food-web patterns in the real world. This promoted confidence that, although they were simple, the models were realistic. But it also led to further insights. The first result, remember, was that species-poor communities are easily invaded, whereas with species-rich communities successful invasion is more difficult. Difficult, but not impossible. If a species-rich community is allowed to mature, it does not remain static but instead experiences a slow turnover of species. In other words, some invasions are successful, usually propelling the loss of existing species; community composition is dynamic, not static. A successful species may, however, become a victim of a later invasion and get pushed out of the community. But its temporary presence leaves a mark on the community, like a footprint in the sand. The second result from Pimm and Post's work, therefore, was that *mature,* species-rich communities are much more difficult to invade than newly established ones. There seems to be something in the maturation process that strengthens the emerging protective network within the community. The community seems to be improving itself, almost purposefully getting better in a way that is difficult to define.

This result is by no means the whimsy of an esoteric computer model, because precisely the same thing happens in nature. In Hawaii, for instance, two types of ecosystem exist. The first is the highland forests, which have been untouched by human interference. They therefore represent a mature, species-rich ecosystem. The second is the lowland forests, which have been disturbed through human activity in and around them. In the process of recovering from such disturbance, they are at an immature stage of assembly, even though they are rich in species. Because the islands have been colonized many times since the first Polynesian settlers arrived fifteen hundred years ago, many alien species came along, too, either as deliberate or accidental passengers. More species of birds and plants, for instance, have been introduced to Hawaii than anywhere else in the world. Twenty-eight percent of the archipelago's insects and 65 percent of its plants are non-native. All its mammals are recent arrivals. Three decades ago, in *The Ecology of Invasions by Animals and Plants,* Charles Elton described the situation there as "one of the great historical convolutions of the world's fauna and flora." Each time an alien species succeeded in establishing itself, it triggered reductions in the population size of native species or it pushed them, cascades of them, to extinction. The issue here is: Where did the alien species become established? Was it in the immature ecosystems of the lowlands or the mature ecosystems of the highlands? Overwhelmingly, the answer is the former. The mature ecosystems were evidently better able to resist invasion than the immature ones. In the terms of ecological theory, the mature ecosystems had reached a persistent state.

Mature communities—whether in the real world or inside computers—clearly have important ecological properties that are denied to immature ones. And the obvious inference is that during the assembly process there is a selection for species that are superior in some ways. Superior, for instance, in productivity, in the case of plant species; in speed or stealth, in the case of predator species; and so on. Self-evidently, a community of superior species will be superior ecologically to one made up of inferior species. However, when Pimm and Post examined the behavioral characteristics of the species in the persistent communities in their computer model, they could find no indication of superiority. In ecological terms, these species were no different from ones that had failed to become part of the community. Perhaps they weren't looking at the right parameters, they speculated.

As it turned out, they had not made a mistake, which became clear when another ecologist, Jim Drake, then of Purdue University, performed similar computer modeling. Like Pimm and Post, Drake promoted assembly of an ecological community by randomly adding species, one by one. But he did it by drawing from a finite pool of species, 125 in all. If a species failed to penetrate on one occasion, it was available for a later attempt. Again, a persistent community matured with about a dozen species. Then Drake started again, using the same pool of species, and again a persistent community matured with about a dozen species. But it was a different community. Less than half the species in the second community were in common with the first. He repeated the process dozens of times, each time getting a mature persistent community, each one different in composition from the others. Again, none of the species in the communities was identifiably "better" in any respect than the others in the pool. Any species could become a member of a persistent community, if it was added at the right time.

These results are as fascinating as they are important. First of all, we can see that persistent communities can form through a process of random addition of species. Second, the ecologically crucial property of persistence, or stability, emerges from the interactions of the species in the community, not through superior qualities of those species. As significant, and perhaps more so, are the implications of these observations on the patchiness of nature. We saw earlier that, traditionally, differences among neighboring ecosystems are explained as a response to differing physical conditions. We saw, too, that chaos theory leads us to expect patchiness, even in the absence of physical differences in the environment. And now, with the work of Pimm, Post, and Drake, we have a second source of patchiness that also does not include adaptation to local conditions: history. The final composition of a persistent ecosystem clearly depends on the order in which species attempt to become part of that system as it matures. Sometimes arriving on the scene early will confer an advantage; sometimes it is better to be late in the day. It all depends on which species are already part of the community. We saw in an earlier chapter that history, or contingency, is becoming recognized as a powerful agency in shaping the path of evolution, while adaptation is seen as having a lesser role than was once assumed. Here we have an analogous situation, with history a powerful agent in shaping the evolution of ecosystems, while adaptation plays a lesser role. That's a very different way of looking at nature from the traditional view.

If it's true of nature, that is. Jim Drake has put it to the test experimentally, by doing with microorganisms (mostly algae of various types) what he does with computer species. He adds the species randomly, and obtains many different persistent communities. History matters. Just recently, two paleontologists produced a fossil-record perspective on this phenomenon. Martin Buzas and Stephen Culver, of the Smithsonian Institution and the Natural History Museum of London, looked at the composition of near-shore marine communities on the North American Atlantic coastal plain over a period of fifty-five million years, during which time the sea level dropped and rose six times. Six times new communities formed in the near-shore habitat, drawn from the pool of species in the region. And six times the composition of the communities was different. Commenting on the observations, the Smithsonian Institution ecologist Jeremy Jackson stated, "This is surely the death knell for the concept of tightly integrated marine ecological communities."[11] It is, and it is also a vote for the importance of historical contingency.

If these results seem counterintuitive, one further observation made by Jim Drake is doubly so. Effectively he said to himself the following: "The persistent communities I make in my computer model clearly work very well. I'll therefore take one of these communities and try to rebuild it from scratch, using only the dozen or so species that make up the community." He could not do it. Once he took the community apart, he couldn't put it back together again, no matter in what order he added the species. Stuart Pimm calls this the Humpty Dumpty effect, for good reason. The explanation is somewhat esoteric mathematically, but it boils down to saying that in order to reach the persistent state, Z, the ecosystem has to pass through states A through Y. You can't jump to Z in one step. If this sounds like the stuff of late-night bar talk at ecology conferences, it's not. These days there is growing interest in restoring ecosystems that have been degraded or destroyed. The prairies of the Midwest and the Florida Everglades are two such examples. In these and other cases, ecologists often know the species composition of the pristine communities, from historical documentation. Until the work I've just described had been done (and it is still being refined), ecologists' inclination was simply to gather the requisite species for the ecosystem they were planning to restore, and then let them loose in the chosen habitat. They were puzzled when they repeatedly discovered it didn't work. Now we know why.

We have seen that nature is not all she seems. There is a dynamic within ecological communities that is counterintuitive and therefore was unsuspected. Communities are always changing, apparently purposefully improving themselves, but we now know that chance and history play a large part. I will finish this chapter with a story of a real ecosystem that reveals these dynamics, shows the importance of change through time, and is a salutary tale for would-be conservationists.

The Chobe National Park in northern Botswana is typical of several ecosystems in southern and eastern Africa. There are many large herbivores, some of them migratory, including giraffe, buffalo, elephant, zebra, wildebeest, and impala. Lions, hyenas, wild dogs, and jackals form a rich carnivore guild. A mosaic habitat of grassland and acacia woodland harbors a rich array of bird and insect species. Altogether, the park offers an abundant diversity of species of the sort that people think of when they hear the word *wildlife*. Managers of the park would like to maintain this diversity, because it is attractive to tourists and because it is perceived of as being the way it should be. They are, however, facing a severe challenge: the acacia woodlands are being destroyed, principally by elephants, and no new trees are growing. If the woodlands shrink to mere remnants of their present selves, the managers believe they will have failed, because they want to keep things as they are. To do so would, however, not only be wrong ecologically; it is also probably impossible. A look at the ecological history of the park reveals why.

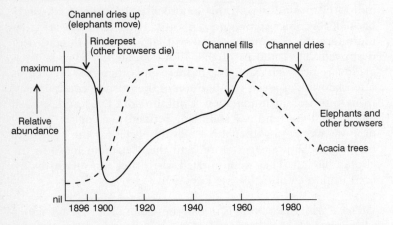

Most environments undergo cycles of change, driven by internal and external forces, as seen here in the history of Chobe National Park, Botswana. (Courtesy of Brian Walker)

The Savuti channel is the major source of surface water in the area. When full, it flows from Angola via the Linyanti swamps and empties into the Savuti marsh (which currently is grassland). It was full in the late 1800s, dried up around the turn of the century, and remained dry until the mid-1950s. In 1982 it dried up again, and remains so. Soon after the channel dried up in the early years of this century, there was a massive outbreak of rinderpest in the area. These two events played midwife to the current acacia woodlands, as follows. The lack of water encouraged the elephants to seek water elsewhere (hunting also reduced their numbers). And the rinderpest epidemic devastated the ungulate population. As a result, browsing pressure in the area was suddenly very light, which allowed acacia seedlings (a favorite food of many browsers) to mature into trees. By the time the elephants and ungulates returned, extensive acacia woodlands had been established. "What we observe today, the coexistence of lots of elephants and extensive *Acacia* woodlands, represents a very narrow window in time and is apparently not sustainable," observes Brian Walker, who has made a detailed study of the region.[12]

It is not sustainable because as long as there are healthy populations of elephants and ungulates in the area, no acacia seedlings will survive to maturity. The animals would have to be removed if the woodlands are to thrive again. "The question," states Walker, "is whether managers and tourists are prepared to accept a ten to fifteen-year period with virtually no animals to see." Probably not. The current species diversity of the park is natural, of course, but it is generated by substantial environmental change that took place over many decades. And change is what park managers often resist; at least they do when they see something of value apparently disappearing. Ecosystems are in a constant state of turmoil, both in space and time, and at any point some populations will be in decline while others may be booming. And constant change is vital as an engine of species diversity. "Conservationists should spend less time worrying about the persistence of particular plant or animal species," warns Walker, "and begin to think instead about maintaining the nature and diversity of ecosystem processes."[13]

Armed with the perspective we've gained about the nature of ecosystems, from an understanding of chaos and the dynamics of the assembly of communities, we can see that what Walker exhorts us to do is sound. But, as with all of human affairs, it is very difficult to manage processes that take many decades to occur. And no one likes to stand idle and watch woodlands shrink or animals die of hunger or thirst. Ultimately, however, that may be what we shall have to do.

10

Human Impacts
of the Past

F OR MUCH of the past two million years, the Earth has
been in the oscillating grip of an Ice Age, with alternating
periods of frigid glacials and warm interglacials. Geologists know
the epoch as the Pleistocene, which ended abruptly between
twelve and ten thousand years ago, giving way to the Holocene
epoch or, simply, the Recent. With each transition between interglacial and glacial, and vice versa, the Earth's plant communities
underwent violent changes. When global temperatures plummeted, tropical forests fragmented and shrank, and forests and
woodlands at high latitudes migrated toward the equator. Because animal species depend on plants for their survival, either
directly or indirectly, they migrated too, if they were able. At the
beginning of interglacials, the reverse process ensued. The Pleistocene generated pulses of global biotic turmoil, in slow motion.
It is no wonder, then, that the epoch witnessed an especially rich
time of the origin and extinction of species.

On the extinction side of the equation, however, there was a
curious imbalance: large species, and particularly large mammals

—those weighing more than a hundred pounds—were particularly vulnerable. Although the largest species that has ever lived —the blue whale—is alive today, the world of terrestrial vertebrates of which we are a part is devoid of such lumbering giants as mammoth and mastodon, Deinotherium and Diprotodon. The so-called megafaunal extinction that produced today's skewed global menagerie is an identifying mark of the Pleistocene, a fact that has puzzled paleontologists for centuries.

But the skewed Pleistocene megafaunal extinctions are a puzzle for reasons beyond the mere fact of them. Outside of Africa, most of the dyings happened late in the epoch, and many at its very end, as the glaciers retreated for the last (for now) time. In the Americas, for instance, fifty species of large mammal slipped into extinction during the past two million years; that is, up to about twelve thousand years ago. Then, in a brief and catastrophic faunal collapse, between twelve and ten thousand years ago, some fifty-seven species of large mammal breathed their last. The coincidence of their demise with the transition between glacial and interglacial times invites an obvious conclusion. Alfred Russel Wallace, codeveloper of the theory of evolution by natural selection, argued in 1876 that the cause of the extinctions rested "in the great and recent physical change known as the 'Glacial Epoch.' "[1] In other words, Wallace believed that the massive disruption of plant communities wrought by the rise in global temperatures was too great for these beasts. They were unable to adapt to the new environments, and therefore died out, he suggested.

Before long, however, Wallace changed his mind, and fingered another potential culprit. In 1911 he wrote, "I am convinced that the rapidity of . . . the extinction of so many large Mammalia is actually due to man's agency."[2] The change of mind was brought about by his growing realization that the environmental effects of glaciation were probably too limited to provoke extinctions on the scale witnessed in the epoch. Wallace wasn't the first to suggest that the end-Pleistocene extinction pulse in the Americas was the work of human hands. In 1860, the British anatomist Sir Richard Owen suggested that the extinctions may have resulted from the "spectral appearance of mankind on a limited tract of land not before inhabited."[3] Earlier still, the Scottish geologist Sir Charles Lyell specifically noted that the extirpation of species through human hunting "is the first idea presented to the mind of almost every naturalist."[4] He was somewhat overstating the case, however, as it did not occur at all to many naturalists. (Earlier than the 1850s, evidence that humans had coexisted with the Pleistocene bestiary was still debated, so an overkill hypothesis was untenable, anyway.) When Wallace embraced the notion of human-induced extinction early in this century, he gave it important support. Nevertheless, the question of climatic versus human impact continued to be debated.

This chapter will examine the impact of humans on ecological communities during recent and not so recent history, usually the result of the colonization of lands where previously no humans had lived. The topic is important for two reasons. First, human colonization of pristine lands is an extreme example of an invading species and the consequences of that invasion on existing communities. We saw that mature, species-rich communities can often resist invasion attempts by most species. But *Homo sapiens* is no ordinary species, and its attempts at invasion are almost always successful and almost always devastating for the existing community. Second, if we are to assess the impact that humans are having on the world of nature today, we need a historical perspective. This chapter gives that perspective.

The ability of the human species to inflict devastation on the natural world at the level of significant extinctions was for a long time thought to be a relatively recent phenomenon in human history. In Wallace's time, biologists recognized that the swaths of European colonizations of the globe from the seventeenth century onward had left a trail of havoc in nature's perceived harmony. Many held earlier colonizers, such as the Polynesians throughout the Pacific, to be blameless in this respect, and to have been part of that harmony. (Western sentiments toward technologically primitive societies had in fact swung dramatically, from their being crude and barbaric beasts to being Rousseauean noble savages.) But as Jared Diamond, a biologist at the University of California at Los Angeles, has pointed out, many pre-European societies felt the same about their own forebears. Two millennia ago, for instance, the Roman poet Ovid wrote: "First came the Golden Age, when men were honest and righteous of their own free will." He was referring not only to the perceived honesty and purity of earlier ages, compared with the treachery and warfare of Greek civilization, but also to their supposed unity with nature.

When Wallace gave support to the idea that human hunters had inflicted mass slaughter on the Americas' menagerie of giant mammals, and, by extrapolation, may have been responsible for Pleistocene extinctions elsewhere in the world, many found it difficult to accept. Some scholars still do. Nevertheless, in recent years it has become undeniable that the evolution of *Homo sapiens* was to imprint a ruinous signature on the rest of the natural world, perhaps right from the beginning. As we will see, there are many means by which we humans make our impact on the world into which we evolved.

The chapter begins with the story of the end-Pleistocene extinctions in the Americas and what can be learned about them by comparison with events in Africa and Australia. It will venture into the more recent history of New Zealand, and identify there strong evidence of human-driven extinctions. And it will describe the fragility of pristine ecosystems of oceanic islands, such as Hawaii, when they were colonized first by early Polynesians and later by Europeans. The message of the complexity of ecosystems —their interconnectedness and their vulnerability to disruption by human hands—repeats again and again.

The Americas of the end-Pleistocene were very different from today's world, so different as to be virtually unimaginable. Almost two thirds of the North American continent groaned under an immense thickness of ice, forming the eastern Laurentide Ice Sheet and the western Cordillerian Ice Sheet, the two separated by an ice-free corridor. A western slither of South America (much of what is now Chile) was also under ice. In between these ice masses thrived a zoo of large mammals, including herbivores such as elephants, mastodons, giant sloths, and lumbering glyptodonts, which carried enormous protective carapaces. These leviathans were prey for lions, giant bears, and saber-tooth tigers, including one known as the smilodon, its eight-inch canines providing the smile of death. There were less exotic creatures, too, such as horses and camels. Within a flicker of geological time, between twelve and ten thousand years ago, these animals were among some fifty-seven similarly large mammal species to go extinct in North America while a much larger number did so in the southern continent. Only a handful of small mammalian species died out, and rats and mice passed through the period unscathed.

The coincidence of this mass dying with the end of the glacial epoch is precise, and would seem compelling as a putative causal agent. Yet there are few detailed hypotheses about exactly how the extinctions might have occurred. It is not sufficient to say that plant communities were plunged into disarray; *therefore* animal species became extinct. This was one of the reasons why, in 1967, Paul Martin, a paleontologist at the University of Arizona, revived the overkill hypothesis of Wallace and Owen, and termed the phenomenon "Pleistocene overkill." He argued that climatic change was not the only event with which the end-Pleistocene extinction coincided. At the same time, a new kind of mammalian species was spreading through the Americas, beginning about 11,500 years ago in the north (after having crossed the exposed Bering land bridge from Asia), and continuing for a millennium, reaching Tierra del Fuego, at the southern tip of South America, 10,500 years ago. That species was *Homo sapiens,* an accomplished hunter, whose predatory skills had been honed for tens of thousands of years in Africa and Eurasia. The immigrants to the New World are known to archaeologists as Clovis people, named after their delicately crafted projectile points, the first examples of which were discovered in 1927 at a site named Clovis, in New Mexico.

Martin calculates that within 350 years of entering North America, the original bands of Clovis people had increased their numbers to 600,000 and had reached the Gulf of Mexico. This explosive expansion was facilitated by unlimited resources—land and prey—opening up before their inexorable advance. Before their first millennium in the New World was over, the Clovis people had reached the southern tip of the continent, and now numbered many million. This north-to-south population expansion left a trail of destruction, as hunters were easily able to kill large, lumbering prey unused to a new kind of predator. The animals probably had no innate fear of humans, as is often the case in regions of the world (usually islands) that have evolved in the absence of humans; they would therefore have been particularly vulnerable to efficient hunters. The hunters, in their turn, were unused to this kind of prey, and so were perhaps freed from the usual hunters' constraint against mass killing. It was, argues Martin, a deadly combination, literally. Evidence of it is to be found in the fossilized bones of the prey species, often associated with Clovis projectile points, throughout North and South America. According to Martin, the direction of the deadly population expansion is to be read in the chronology of the fossil sites, the oldest in the north, the youngest in the south. Not all paleontologists agree that the pattern is this clear, however.

The deinotherium, one of the giants of the Pleistocene that became extinct at the hand of man.

Before I go on to discuss the merits of the overkill hypothesis, I should say that it is part of one of the more hotly debated issues of the anthropology of recent human history. That is, when did the peopling of the Americas occur? The discovery of the first

The glyptodont, another of the giants of the Pleistocene that became extinct at the hand of man.

Clovis points early this century, followed by hundreds of similar finds in subsequent decades, seemed to be signs of the entry event. When radiocarbon dating on material associated with Clovis points gave an age of 11,500 years, the question appeared settled. Nevertheless, prior to the unearthing of the Clovis site, and persisting after, were claims for signs of human habitation far earlier than 11,500 years, some as much as 35,000 years ago. Evidence for some, but by no means all, of these pre-Clovis sites looks convincing. Even so, there are few such sites, and it is possible to imagine a series of migrations into North America when glaciation lowered sea levels sufficiently to expose the Bering land bridge that joins Alaska with Siberia. Pre-Clovis entry may have been sparse; in any case, the archaeological imprint implies that significant population growth did not result from them. Only with the coming of the Clovis people does the evidence suggest rapid population expansion, in numbers and in territory occupied. Whatever the date of the *first* entry, it does not detract from the overkill hypothesis linked to the end-Pleistocene expansion of the Clovis people.

One of Martin's arguments in support of his hypothesis is the uniqueness of the event. "If Ice Age climatic changes were important in determining the extinction of American large mammals," he wrote recently, "it is not obvious why earlier glaciations

The woolly mammoth, a third giant of the Pleistocene that became extinct at the hand of man.

and interglacial warmups were unaccompanied by faunal losses.''[5] The two million years prior to the end-Pleistocene faunal collapse were punctuated by many glacial-interglacial transitions, yet no major extinction pulse is associated with any of them. And why, Martin asks, if there was something particularly deadly about the end-Pleistocene warm-up, did a similar pulse of extinction not occur in other parts of the world, such as Australia and Africa? Moreover, if climate was the culprit in the Americas, through the devastation of plant communities upon which the animals depended, how do we explain that the plant species that were important in the diets of, for instance, mammoth and ground sloths remained abundant and widespread after these mammalian species became extinct? These are cogent points in favor of the overkill hypothesis.

Nevertheless, there is another side to the argument. Why, critics of overkill ask, if hunters wiped out so many large mammalian species, did some species survive, such as the bison, moose, elk, and musk ox? If the impact of hunting was so devastating, why were these species spared? Perhaps the answer lies in the special history and behavior of these animals. "Many of the American megafaunal survivors were themselves immigrants from the Old World, where," observes Martin, "despite a much longer history of being hunted and despite severe reductions in range, they also managed to survive."[6] Having coevolved with humans for several million years, during which hunting skills developed, these Old World species had acquired a wariness of humans and instincts for avoiding their predatory habits. These species took their life-saving characteristics with them when they migrated from Asia into the Americas. Some of the survivors were unpredictable in their movements, too, such as caribou, bison, and antelope, while others, such as moose, spectacled bear, and tapir, lived in environments of heavy cover, where detection was more difficult. Whatever the reason, survive they did, but they represented only a minority of large mammalian species of the end-Pleistocene fauna.

Some of the supporters of the climate hypothesis argue that it is often portrayed too simplistically. A plunge in global temperature does not just shrink and fragment plant communities, leaving remnants where reduced populations of animal species might survive on them. Instead, the integrity of the communities is shattered, with some species migrating to one kind of new habitat and others to different ones. The result is the formation of communities that did not exist prior to the climate change. Russell Graham and Ernest Lundelius, biologists at the Illinois State Museum and the University of Texas respectively, call this phenomenon "coevolutionary disequilibrium." At a major conference on the topic a decade ago, Graham explained the model as follows: "Late Pleistocene communities are characterized by the coexistence of species that are now geographically and ecologically separated. This implies that communities did not migrate as intact units but instead responded to environmental change in accordance with their own tolerance limits. Significant adjustments in feeding strategies were required by many animal species." It may be true, concede Graham and Lundelius, that some of the plant species on which the ground sloth subsisted are still to be found in the region where these animals slipped into extinction; but they almost certainly required other plants for their subsistence, too. Perhaps these did disappear from the sloth's habitat. Unfortunately, it is not easy to test the proposal.

There is, of course, a middle ground in this debate, one that combines the deadly impact of hunting and the climate change. This is the scenario proposed by John Guilday, a specialist in the plant and animal life of the Appalachian Mountain region. A firm believer that the drastic climate change of the end-Pleistocene could not fail to have had a devastating effect on the populations of animal species, he nevertheless accepted that hunting could have played a part, too. "It is possible to visualize a primitive hunting overload that could result in the extinction of prey species," he wrote shortly before his premature death in 1982, "but only under conditions not directly related to hunting pressure per se, but rather to the overall ecological health of that species."[7] It would, he suggested, be folly to insist that hunting alone was the culprit. "Large mammal taxa did not simply vanish from a scene of ecological composure, but instead disappeared during a time of great ecological ferment when biotas were adjusting, dissolving, and reforming under new climatic parameters to emerge into the Holocene in greatly altered aspect."[8] Established ecological niches disappeared; new ones were established. The issue is: How dependent were the doomed large mammals on the specific composition of the ecological communities with which they coevolved? Great dependence would have left them vulnerable to change; little dependence would have allowed them to shift their subsistence strategies to the newly established communities. We do not know the answer.

Currently, there is no way of teasing out the relative impacts of hunting and of climate change on the end-Pleistocene fauna of the Americas. Both events were visited simultaneously on the continent. Did one process so weaken the fauna, through reduction in population size, that the second process was sufficiently detrimental to finish them off? As Guilday put it, "In any event, their combined effect was devastating, and the world is much the poorer."[9]

• • • •

Europe was once home to many large mammalian species, but is no longer.

I have already stated that the pattern of Pleistocene extinctions in Africa and Australia is different from that in the Americas. There was no dramatic faunal collapse at the end of the epoch, twelve to ten thousand years ago, as there was in America. Africa and Australia did not escape culling of species during the Pleistocene, however. In fact, Australia suffered proportionately even more than the Americas, losing some 85 percent of its large mammalian species. Africa fared much better, but nevertheless experienced significant extinctions early in the Pleistocene. It remains, however, the continent of big game par excellence. The diachrony of Pleistocene extinctions in Africa, Australia, and the Americas surely undercuts the argument for climate change as the culprit in the Americas. Climate change is global, and would be expected to exert similar effects in all continents at the same time. The diachrony may also be used in favor of the overkill hypothesis, as follows.

First, the extinction event in the Americas coincides closely with the spread of Clovis people there, as we've seen. What of Africa, which experienced important extinctions early in the Pleistocene, and Australia, where extinctions were late, perhaps sixty thousand years ago? A coincidence of the arrival of humans and significant extinction pulses would support the overkill hypothesis.

Africa is the cradle of humankind, as Charles Darwin noted more than a century ago. The first members of the human family evolved there some five million years earlier, but the subsistence patterns of these early human species was much like that of apes. We were not hunters from the very beginning of our evolutionary career. If we are to be guided by archeological evidence, meat-eating became important sometime between two and three million years ago, coincident with brain expansion. With the evolution of *Homo erectus* almost two million years ago, a hunting-and-gathering mode of existence had become well developed. It continued to be so until very recently, when agriculture was invented, about ten thousand years ago. I'm not suggesting that *Homo erectus* was as accomplished a hunter as modern technologically primitive people are. Indeed, Richard Klein, an anthropologist at Stanford University, has shown convincingly that hunting skills improved considerably with the evolution of modern humans, about 100,000 years ago. Before this, people selected docile, easy-to-catch prey; afterward, even dangerous animals, such as Cape buffalo, became common prey. As emergent hunters, members of the genus *Homo* coevolved with potential prey for more than two million years.

Paul Martin has argued that, during this long coexistence, animals evolved the means of avoiding the predatory habits of humans. "The distinction between African and American Pleistocene extinctions is seen in the difference between gradually developing [humans] evolving for millions of years with large animals on one continent, compared with the onslaught of a highly advanced hunting society at the height of its power suddenly arriving on the other," he states. "Had America rather than the Old World been the center of human origins, the late Pleistocene record of extinction might well have been reversed."[10] Were the megafaunal extinctions early in the Pleistocene the result of an initial impact of a new predator *(Homo erectus)*, to whose presence potential prey species had not yet adapted? Such a coincidence would be consistent with the overkill hypothesis. But there is no way to determine whether this is true, and it is just as likely that climate change was the cause.

Australia offers a more clear-cut message, albeit not without some uncertainties. By virtue of its long geological isolation, its native fauna of the late Pleistocene was unlike any other continent. There were rhinolike creatures, ground sloths, giant kangaroos, tapirlike animals, giant capybaras, but only two carnivores, a lion and a dog—all, of course, were marsupial mammals, not placental mammals. (There was a third large carnivore, a giant lizard, larger than the Komodo dragon.) Unlike the Americas, Australia lacked an elephant-sized mammal and giant ground sloths. Proportionately, the end-Pleistocene extinction in Australia wiped out more than the three-quarters decimation that occurred in the Americas. (The total number of species was, however, less.) Of the fifty species of megafauna that existed prior to sixty thousand years ago, only four survived, all kangaroos. As in the Americas, very few small mammals became extinct.

For various reasons, it is difficult to determine a secure age for the fossil deposits of Australia's extinct mammals. The best estimates put them at between 100,000 and twelve thousand years ago, probably about sixty thousand. This is very close to the date of human entry into the continent, as estimated from archeological evidence. Does this coincidence point an accusatory finger at *Homo sapiens,* indicted again as perpetrator of continentwide carnage? Paul Martin believes it does. "The arrival of a potent and deadly species, *Homo sapiens,* landing on a continent that had previously known few large cursorial carnivores, and none of these in the order Carnivora, seems uniquely favorable for overkill," he states. "The loss of the less fleet and more conspicuous large mammals seems inevitable and unremarkable."[11]

If that is so, respond critics of the overkill hypothesis, where is the evidence? Where are the petrified skeletons of giant kangaroos impaled on stone spears? The virtual absence of such associations of carcasses and ancient weapons is surely a weak point in the overkill hypothesis. Martin explains it by suggesting that, given the short period of time between human entry, subsequent population expansion, and the extirpation of species, virtual archeological invisibility is not surprising. It is true that the fossil and archeological records capture slow-moving images, not fast action, so Martin's argument is valid. Nevertheless, it is also untestable scientifically, and that makes scholars uncomfortable.

Evidence of detrimental impact by humans on ecological communities into which they insert themselves is therefore principally circumstantial. Whether the Pleistocene extinctions were caused by human agency or not, the devastations in these communities in relatively recent times illustrate their fragility. Change some external circumstance, and they might collapse.

If there is a question over what dispatched the giant denizens of the Americas, Africa, and Australia, no such uncertainty exists about New Zealand. An isolated neighbor of Australia in the Southern Hemisphere, the two islands that constitute New Zealand were until recently also host to a unique biota, one "so strange that we would dismiss them as science-fiction fantasies if we did not have their fossilized bones to convince us of their former existence," as Jared Diamond has written.[12] It was a land of birds, but of the most extraordinary kinds, many of them flightless. The stars on this stage were the giant moas, ostrichlike creatures that stood ten feet tall and weighed 530 pounds. A dozen moa species thrived, the smallest of which stood a mere three feet tall.

The evolutionary experiment that played itself out on New Zealand through millions of years of its prehistory is a study in ecological opportunity. Mammals were all but absent (there were bats), so the birds and other creatures filled mammal-like niches. Diamond describes it this way: "Moas instead of deer, flightless geese and coots instead of rabbits, big crickets and little songbirds and bats instead of mice, and colossal eagles instead of leopards. The scene was as close as we shall ever get to what we might see if we could reach another fertile planet on which life had evolved."[13] The eagle, which weighed about thirty pounds, was the most powerful aerial predator of its time, and was the moas' only enemy. That is, until humans arrived.

The islands were first inhabited, almost a millennium ago, by Maoris, as the Polynesian settlers were known. The world they found awaiting them was as I've just described. Within a few centuries that world was no more, transformed by massive local extinction. Eventually almost 50 percent of the islands' species vanished, including all the large birds and most of the flightless species, too. Until very recently, most observers believed that the Maori people were careful conservationists and had no part in the devastation of the islands' biota. Climate change was the cited culprit, something that was assumed to have occurred prior to the Maoris' arrival. It was conceded that perhaps the Polynesian settlers found the tattered remnants of a once fecund environment, and played a hand in finishing them off. But humans were assumed to have been only bit players in this scene of destruction.

Two recently assembled lines of evidence have undercut this belief. First, it is clear that the New Zealand climate since the end of the Pleistocene has been equable and favorable for the avian-dominated life there. "The moas died with their gizzards full of food, and enjoying the best climate that they had seen for tens of thousands of years," notes Diamond.[14] There seems to be no suggestion here that the moas were eking out a meager existence in an impoverished environment. Second, the litter of Maori living sites reveals clearly that the moas were still thriving when the first humans arrived. The bones prove it. Moreover, the evidence is not sparse, but voluminous. More than a hundred such sites are known, some of them extremely big. Moa remains show that the Maoris exploited the birds for food, which they cooked in earth ovens, and for material, such as skins for clothing and bone for fishhooks and jewelry. Blown-out eggs served as water containers. As many as half a million moa skeletons have been recovered from archeological sites so far, about ten times as many individuals as would have been alive at any one time. The Maoris must have been slaughtering moas for many generations before the birds became extinct.

One would have guessed that moas, given their size, muscle power, and potential speed, would have been formidable adversaries, even for skillful hunters. And, given the extremely difficult, mountainous terrain of New Zealand, they would not have been easy to track down and corner. Or would they? Perhaps because moas and the other New Zealand natives had evolved in the absence of humans, they were very tame. Getting close enough to dispatch them might not have been difficult at all. But hunting is not the only means by which the islands' species suffered at human hands. After all, in the company of the moas as they slipped into extinction were, among other species, the islands' extraordinary crickets, snails, wrens, and bats. These are not typical prey species. Very probably, deforestation was an important factor here. As the Maoris cleared the land for their settlements, they eliminated habitats on which the species depended. The Maoris also brought rats with them, which had a devastating impact on ground-living birds and others species. The nests of these creatures offered easy pickings to this most voracious and versatile predator. New Zealand was not the first nor the last land to suffer massive ecological depredation following the introduction of rats. Rats have no trouble invading an ecosystem, and the result is usually a cascade of extinction.

With the moa populations first in decline and then finally absent, the giant eagles would have had no prey on which to subsist. They may well have tried to substitute humans as prey, because, like moas, they are bipedal animals of similar height. The Maoris would have taken steps to protect themselves, and there was only one outcome in that confrontation. The notion of man-the-exterminator is secure in New Zealand. Even proponents of climate-change hypotheses, like John Guilday, concede as much. He described the case as "clear cut." When Europeans arrived in New Zealand in the 1800s, they found a people apparently in harmony with their environment, which initially was imagined to be pristine, a glimpse of an earlier world. Before long, however, the European colonists' plows began to unearth the bones and eggshells of exotic birds, the like of which had never been seen before. It was the first glimpse of an important story about ourselves—*Homo sapiens*—that has only recently begun to unfold. The fact that the first presence of humans in New Zealand was devastating to its existing ecological communities adds circumstantial support for similar impacts in the Americas.

Oceanic islands like New Zealand are eccentric corners of the evolutionary process, for the very reason of their isolation. They become populated by whatever species chance—and physical circumstance—brings along, and by the descendants of those species. With the exception of bats, mammals are typically absent from distant islands, which makes the ecological communities

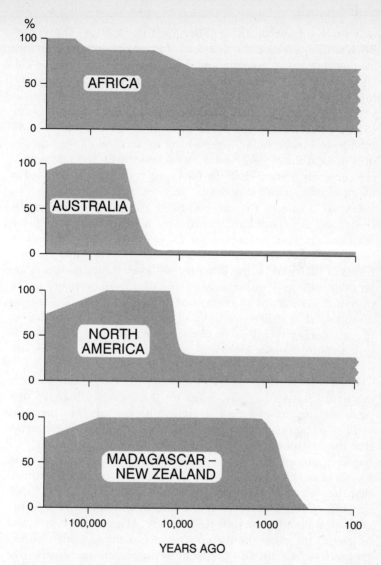

%

100

50

0

AFRICA

100

50

0

AUSTRALIA

100

50

0

NORTH
AMERICA

100

50

0

MADAGASCAR –
NEW ZEALAND

100,000   10,000   1000   100

YEARS AGO

Many lands suffered loss of large animal species during the Pleistocene, but at different times. With Australia, North America, New Zealand, and Madagascar, the losses coincided in time with colonization by humans. The extinctions were caused by a combination of overhunting and destruction of habitat.

that emerge there distinctly different from those on continents. Birds and reptiles are the denizens of these communities, so they frequently have an aspect about them of being of another world. Island ecological communities are also extremely fragile and vulnerable to devastation following invasion by alien species, particularly mammals, whether humans or rats (which usually go together). For instance, only 20 percent of bird species live on islands, but more than 90 percent of extinctions of bird species in historical times were island forms. Moreover, oceanic islands are currently home to half the bird species that are recognized as being in danger of extinction in the near future.

Hawaii is one of the most isolated groups of islands in the world, and as a result harbors many plant and animal species that live nowhere else. It is a paradise for ecologists and evolutionary biologists alike, who find ways to glimpse the fundamental processes of life there, often in a pristine state. It has, however, suffered enormous devastation as a result of human occupation. For example, as much as 70 percent of its population of bird species disappeared as a direct or indirect result of human presence on the archipelago. Only a small percentage of its forests are untouched, those in the highlands, where the tentacles of economic development are unable to reach.

Until very recently, scholars assumed without question that the ecological havoc that was visited on the islands followed European colonization, in the late eighteenth century. For instance, just two decades ago a biologist at the University of Hawaii wrote that the "serious degradation of the Hawaiian environment . . . began in earnest a few years after the arrival of Captain Cook and his successors."[15] Just as with New Zealand, scholars assumed that the Polynesian settlers of Hawaii, who reached the archipelago fifteen hundred years ago, established a harmony with the ecological community they found there. Hawaiians were viewed as part of the natural environment, just as the Hawaiian honeycreepers were. Captain Cook and his immediate successors were assumed to have set eyes on a pristine ecosystem, one that had recently embraced the addition of an innocuous mammalian species, *Homo sapiens,* in the form of the Polynesians.

In the 1970s, this view began to be undermined, as, first, Joan Aidem, an amateur naturalist on Molokai Island, and then Storrs

Olson and Helen James, of the Smithsonian, started to piece together earlier life on Hawaii. Working with fragile, fossil bird bones, they saw that the ecological communities of the islands were very different from what had been imagined, and that much of what had made them so different was extinguished within a few centuries of the arrival of the first Polynesian settlers. "We had long known that the islands were once clothed in virgin forests composed mostly of plants found nowhere else in the world," they wrote after five fruitful years of exploration, "but only in the last twelve years have we begun to realize the extent to which the forests were home to a remarkable variety of endemic species of birds, most of them doomed to rapid extinction at the hands of humans."[16] As many as fifty species of birds became extinct in the prehistoric Polynesian period, including ibises, geese, rails, owls, a hawk, an eagle, ravens, and many, many song birds.

When Olson and James were poring over the fossil bird bones, drawing up the list of the dead, they were unprepared for the high proportion of flightless species: some seventeen of the fifty. "The existence of these unusual birds was Hawaii's best kept secret up to the last decade," James and Olson wrote.[17] The extinct birds were not members of obscure lineages, long consigned to extinction, but were descendants of familiar species, such as ducks, geese, rails, and ibises. Their ancestors must have been accomplished fliers; otherwise they would not have been able to make the 2000-mile journey from the nearest continental land mass to the distant archipelago. Released from the lethal attentions of the predators with which they had coevolved, many of these species gave up the ability to fly, an evolutionary transformation that, biologists have come to recognize, often happens on islands. Given the opportunity not to have to expend the energy to fly, many species do just that. It was a relatively rapid process in evolutionary terms, because the oldest of the main Hawaiian islands, Kauai, was thrust through the waves as a biologically virgin volcano just six million years ago. The rate of disappearance of these birds was much greater, measured in centuries or even decades.

"It seemed reasonable to conclude that the Hawaiian avifauna was in natural equilibrium when the islands were first visited by Europeans," noted Olson and James, referring to the received wisdom they helped overturn. "The fossil record, however, has shown that the historically known avifauna of the archipelago constitutes only a fraction of the natural species diversity of the islands."[18] On Oahu, for instance, seven species of bird are known historically, and yet in fossil deposits there are relics of four times this many. This was a tremendous impoverishment, again coinciding with the arrival of Polynesian settlers. Further research has shown that the prehistoric fate of the Hawaiian Islands was not exceptional.

In other regions of the Pacific, the two Smithsonian researchers and several colleagues discovered that wherever they were able to find fossil evidence of bird species on oceanic islands, they could draw two conclusions: first, that the present ecological communities are disastrously impoverished compared with prehistoric times; second, that the extinction of the species always coincided with the arrival of the first settlers. The example of Henderson Island is extreme, and illustrates the general point. Located in the Pitcairn group of Pacific islands, Henderson is small (thirty-seven square kilometers) and not particularly inviting. Steep limestone cliffs rise from deep water, the terrain is uneven and densely vegetated, and sources of fresh water are scarce. Many of the animal and plant species there are found nowhere else. Not long ago it was described as "one of the few islands of its size in the warmer parts of the world still little affected by human activity." As the island was considered to be in a relatively pristine state, biologists felt confident that they could record its species, and infer reliable information about the structure and interactions of an intact, fully functional community.

They were wrong to do so. Joined by David Steadman of the New York State Museum, Olson explored the island's fossil deposits and uncovered a very different picture. Rather than being pristine, Henderson lacks at least one third of its land bird species compared with prehistoric times, and probably more. Henderson was occupied briefly by Polynesian settlers sometime between eight hundred and five hundred years ago, and has come to be called a "mystery island," because no one knows why the inhabitants left. There are a dozen or more such mystery islands in the Pacific, but in the case of Henderson, the mystery is why the settlers stayed as long as they did, given the grinding subsistence conditions they must have faced. There is, however, no mystery anymore to the list of extinct species to be found on these islands. They were extirpated as a consequence of human presence.

Although the hunting of bird species undoubtedly drove some of them to extinction, there are other means by which human presence detrimentally affects natural ecosystems. And whereas hunting may affect one or a few species only, other avenues of impact can be more far-reaching. In Hawaii, for example, settlers cleared much of the lowland forest, an activity that destroys and fragments habitats. Cut down a forest, and the organisms living there must find similar habitats, or perish. Cut a swath through a forest, and communication and travel across what used to be continuous habitat may be hindered or blocked completely. Ecological communities may be disrupted, making them more susceptible to invasion by new species, perhaps precipitating cascades of extinctions. A recent fascinating study by British and American ecologists reveals that some of these extinctions were not only unexpected but also were long delayed following the initial disturbance.

Habitats that are disturbed through fragmentation lose species for several reasons. Those which require very large foraging ranges, for instance, will disappear from fragmented habitat; this particularly affects top carnivores. Species that have low population numbers are vulnerable to extinction through chance events. And so on. The vulnerability of these doomed species is readily identified. But when David Tilman, Robert May, and their colleagues from the Universities of Minnesota and Oxford carried out mathematical modeling on complex communities, they found a result no one had predicted. Among the most vulnerable species in such communities are those which are best adapted. But that vulnerability may take decades or even centuries before becoming apparent. I'll explain.

Think of ecological communities as being composed of two types of plant species, those well adapted to local conditions, and those less so. The well-adapted species expend their energy in exploiting resources in their habitat, through having deep roots to reach a low water table or resist periodic fire, for instance; or having tall stems so as to harvest light more readily than shorter species. The strategy adopted by the less well-adapted species is to be more mobile, through seed dispersal. This allows them to take advantage of fleeting opportunities for colonizing new habitats. When such an ecological community becomes isolated, through habitat fragmentation, the species most at risk are those which are least mobile; that is, the well-adapted species. Trapped in isolated patches, these small, local populations become vulnerable to occasional catastrophes, such as disease, fire, or a shortage of nutrients. One by one, the isolated populations become locally extinct, until eventually they disappear from very large regions or vanish completely in a slow march to oblivion. "Because extinctions occur generations after fragmentation," noted Tilman and his colleagues, "they are a 'debt' that becomes due in the future."[19] This work implies that the effects of habitat disturbance as much as half a millennium old may still be rippling through today's world. Similarly, the havoc of current environmental destruction is building up extinction debts that will be visited on our children's children, five hundred years hence.

Destruction and fragmentation of habitat are now recognized as being responsible for much of the extinction pulse suffered in Hawaii. But humans seldom travel alone in their settlement quests. Some creatures they take along intentionally, such as cats, dogs, pigs, and goats; others are unintentional stowaways, such as rats. Introduce these creatures to island communities that are innocent of previous mammalian presence, and mayhem ensues, either through competition or predation. Goats are browsers par excellence, and easily outchew native species, threatening to denude natural landscape in the process. Goats were introduced onto several of the islands in the Galápagos archipelago two centuries ago, and reduced some of them to virtually bare rock. But predators have a more immediate impact, and they don't have to be large and aggressive to be extremely dangerous to the life of a community. It may come as a surprise to many to learn that of all the predatory fellow travelers of humans, rats are responsible for the largest number of extinctions. With omnivorous diets, rats dine on the eggs and the young of birds and reptiles, beginning the chain of destruction early in the life cycle.

As we saw, ecological communities are not simply collections of species living by happenstance in the same location. They are subject to interactions, albeit weak, through complex food chains. As a result, the extinction of a single species may reverberate throughout the community, producing torrents of further extinctions. For instance, several Hawaiian plant species are on

Once the most common bird in America, the passenger pigeon is now extinct. A natural history book published in 1874 described the bird as follows: "The passenger pigeons traverse the forest of America in such compact masses that they absolutely intercept the rays of the sun, and cast a long track of shadows on the ground . . . The passing of these columns sometimes lasts three hours." (Some of these columns contained as many as a billion birds.) And yet less than twenty years after these words were written, the passenger pigeon was extinct in the wild, the victim of massive hunting. As this same naturalist noted: "Scarcely are the pigeons installed [in the forest] than all the able-bodied people in the country hasten to the spot, and make a complete carnage of them."

the verge of extinction because of the disappearance of honeycreepers, upon whose long, curved beaks the flowers depended for pollination. This is an example of direct dependence; but chains of dependence can be indirect, too. Large grazers often make it possible for smaller grazers to live where they might not be able to otherwise, because the effect of the large species' activities is to open up the habitats. The disturbance effects of these megaherbivores promote diversity and greater productivity among plants in the habitat. An example of what happens when such "keystone herbivores" are removed (that is, become locally extinct) is seen in the disappearance of elephants from the Hluhluwe Game Reserve in Natal. Within a century, three species of antelope became locally extinct, and even populations of wildebeest and waterbuck were reduced. When the elephants disappeared, so did their habit of generating habitats for lesser herbivores.

Ecological communities are complex systems, as we saw. The practical consequences of that complexity is evident in the impact of disturbance on those communities. By now it is undeniable that an important component of that disturbance in the recent past has been shaped by the human presence. The scale of ecological devastation, particularly on oceanic islands during the past few millennia, has only recently been appreciated, and it has come as a sobering realization. Not only is much of the world not as ecologists imagined—that is, it is not pristine, revealing natural systems in intact assemblies—but we cannot escape the fact that the human species has had a detrimental impact on the global ecosystem into which it evolved so very recently.

It is not necessary to employ machines of mass deforestation to perpetrate great environmental abuse. Technologically primitive societies have chalked up quite a record in this respect in the recent past, precipitating what Storrs Olson has called "one of the swiftest and most profound biological catastrophes in the history of the earth."[20] Add to this the possible mass exterminations in Australia and the Americas, and *Homo sapiens* holds claim to a long history as an agent of extinction. We now inhabit a modern world, with this history behind us. In the final section of this book I will address the present and future impact our species is threatening on the world. Some disagree, but I believe that what we face is profoundly disturbing, both in the loss of magnificent individual species, such as elephants, and in the global effect. I have talked here about human impacts of the past, not as an excuse for what is being visited now. We are aware of what we are doing and its consequences; earlier societies were not. It is, however, important to view the patterns of the present in the correct historical perspective.

11

The Modern Elephant Story

ELEPHANTS are the largest living terrestrial animal. Majestic in form and movement, complex and preternaturally sensitive in behavior, they have given rise to much mythology. In the Hindu religion in India, there is an elephant-faced god, named Ganesha, who is all-powerful and can overcome any obstacle. Ganesha, who is also the patron of literature and learning, is invoked at the start of worship or at the beginning of challenging and uncertain endeavors. Hindu legend has it that the elephant once had the power of flight, but lost it when, one day, it landed on a banyan tree and fell onto a hermit's house, destroying it. The hermit's curse doomed it to remain on terra firma, deprived forever of flight. To the Romans, elephants were inscrutable creatures that worshiped the sun, moon, and stars. Aristotle described them as being "as near an approach to man as matter can approach spirit." Pliny the Elder, the great Roman historian, concurred, allowing that, in all of the animal kingdom, these magnificent pachyderms were "closest to man." Elephants were held to understand human language, to have a moral and ethical sense, and to elaborate a culture that, though different from that of humans, was as intricate and spiritual.

These days, elephants are seen to epitomize "the wild"—they are powerful and free, intelligent and, yes, still enigmatic. They are nature's splendid creation, a tangible link with a deep and hidden past, and unchallenged masters of the plains. If you go to Africa to see the quintessence of wildlife, you go to see elephants.

During my five years as director of the Kenya Wildlife Service, I was preoccupied with elephants. This was not for esthetic or academic reasons but for the gruesomely compelling reality that they were hurtling toward extinction. Deprived increasingly of their once endless range by inexorable growth of human settlements, and slaughtered in cold blood for their ivory, the elephant population of Africa had been halved in the decade that led up to my appointment as director, in April 1989. Unchecked, these twin forces of destruction were on course to dispatch the elephant to evolutionary oblivion by the turn of the century. Beginning the new millennium with the blood of so glorious a species on our hands would have been sickening testimony to human irresponsibility and greed. This appeared to matter not at all to some; but it mattered to me.

Fortunately, the plunge toward extinction has been halted, an achievement in which I am proud to have played a part. There is no room for complacency, however. Stopping the slaughter was just the first step along a difficult and uncertain path. Whether there will be elephants in the wild to inspire awe in our children's children, and in their children, just as they do in us today, depends on how that path is negotiated. I will describe something of the circumstances that pushed elephants on a course toward extinction, and how those circumstances were changed. But I will not dwell in detail on this recent episode, with its mix of conservation biology, politics, and trade. For that story, readers can do no better than to turn to *Battle for the Elephants,* a recent book by Iain and Oria Douglas-Hamilton, two dedicated and tireless people I was lucky to have as friends and allies in a joint struggle. Here, I will concentrate more on the lessons we can learn from elephants—from their history, present plight, and uncharted future.

In many ways, elephants and their place in the world encapsulate the central themes of this book: namely, that through time we see change; and through change we see the generation of diversity in nature. Not surprisingly, being the largest terrestrial species of animal, elephants face extreme challenges as far as coexistence with humans is concerned. Negotiating the path to that long-term coexistence is, as I said, uncertain, and we may want to appeal to Ganesha for guidance as we embark on the venture.

I'll begin by putting elephants into a broader biological context, seeing them as a modern species with a deep evolutionary past. I'll then recount the more recent history of the species, which essentially is its relationship to—or perhaps I should say exploitation by—humans. This will bring us through to the present, where the possibility of anthropogenic extinction is very real. Finally, as a way of understanding the ecological and biodiversity issues associated with saving the elephants, I'll describe the challenges faced by a population of these wonderful creatures in Amboseli National Park, located in southern Kenya, at the foot of Kilimanjaro. In many ways, the elephants' story is the story of biodiversity in the world, through time and space.

There are two species of elephant in the modern world, the African elephant, with the biological name *Loxodonta africana,* and the Asian elephant, misleadingly named *Elephas maximus,* because it is the smaller of the two. There are actually two variants, or subspecies, of the African elephant: *Loxodonta africana africana* is the one seen by most visitors to the continent, as it is the denizen of the woodland savannah of East Africa; the second, *Loxodonta africana cyclotis,* is smaller and lives in the forests of Central and West Africa. This small coterie of modern leviathans is in fact the remnant of what, over the past fifty-five million years or so, was a dominant group, or order, of large mammals, known as the Proboscidea. The choice of this ordinal name, which derives from the Latin word for nose, *proboscis,* is apt, because the elephant's prehensile trunk is unique among mammals. Get close to an elephant—a tame one, I'd suggest—and you'll quickly discover how versatile an organ the trunk is. More than just an extra, deft limb for harvesting food or wresting obstacles out of the way, it is the means by which the animal knows intimately about its world, seeking scents drifting in the air or lurking in the folds of the body.

Whenever discussion turns to the subject of elephants' closest living relatives, reaction among nonbiologists is usually one of incredulity: sea cows (that is, manatees and dugongs). The only truly aquatic herbivores among mammals, the sea cows' claim to evolutionary proximity to modern elephants rests not on their large size, but more on the anatomy of their bones and teeth. They also have mammary glands on their chests, as elephants do, not on their abdomens, as is the case with most mammals. Biologists acknowledge these similarities by classifying sea cows and the Proboscidea into a mega-group, or superorder, named Subungulata.

The origin of the Proboscidea is shrouded in mystery, but we do know that they were part of the burst of evolutionary activity that followed the end of the Cretaceous period, sixty-five million years ago, when the reign of the dinosaurs was terminated by Earth's collision with a giant comet or asteroid. Earth's biodiversity was drastically reduced during that mass extinction event, the last of the Big Five. The evolution of the Proboscidea was part of the surviving biota's inevitable response in a rapid rebuilding of diversity, but, as always, with many novel players entering the scene.

The earliest known proboscidean has been found in southern Algeria. It is the fossilized skull of a swamp-living animal from the early Miocene epoch, some fifty-four million years ago. The beast apparently stood less than three feet tall, but the anatomy of its head reveals the characteristics of a once present prehensile trunk, hallmark of all proboscideans. Biologists love to speculate on the origin of the prehensile trunk. Under what circumstances might such a structure be advantageous? The ecological setting of some later proboscidean finds indicates they were aquatic or semiaquatic creatures, like the Algerian species. Perhaps this is a vital clue. Perhaps, speculation runs, natural selection favored the evolution of an organ that could harvest vegetation while the animal was in shallow water. A rudimentary trunk, which is formed from the lips, palate, and nostrils, could do that job. True or not, the trunk became an important organ for proboscideans, and its evolution was eventually accompanied by the evolution of tusks. Anyone who has seen an elephant uproot a tree knows what a potent combination trunk and tusks are in accomplishing that job. The tusks are also important in displays of rivalry among males. (The tusks are considerably larger in males than in females, which is not surprising, given their difference in body size.)

Fully half of proboscidean history was played out in Africa, during which time several major subgroups appeared. Beginning about twenty million years ago, descendants of these groups eventually expanded into every major landmass in the world, with the exception of Antarctica and Australia. The head count of species in the Proboscidea across the globe at that point in prehistory now stands at close to two hundred. It was a diaspora of an extraordinarily successful brand of animal, and the group dominated the Age of Mammals. The fact that only two species remain in today's world is yet another reminder that dominance is not forever: it never has been in the history of life, and it never will be. We should take note.

Many of the animals that make up the major proboscidean groups are all too familiar to us, having been so often depicted in dramatized scenes from the Ice Age: they include mammoth and mastodon, which were striking members of the fauna of North America and Eurasia. Their images are also to be seen in cave paintings in Europe, from twenty-five thousand years ago. Others are less familiar, such as gomphotheres, lumbering giants equipped with tusks formed from broad lower incisor teeth, which made it all the way to South America. They had upper tusks, too. Another group, deinotheres, also had lower tusks, but, unlike those of the gomphotheres, these curved downward. Again unlike gomphotheres, the deinotheres lacked upper tusks. We have uncovered several splendid specimens of deinotheres in the hominid-bearing deposits around Lake Turkana. The immediate ancestors of the modern elephant species evolved before the onset of the Pleistocene Ice Age, beginning two million years ago. The Asian elephant, *Elephas,* probably arose first in Africa and then spread into much of the rest of the Old World. *Loxodonta,* the African elephant, evolved first in the dense forests of that continent, sometime in the Pliocene epoch, between five and two million years ago. It is a matter of some irony that the immediate ancestors of the modern elephant evolved roughly coincidentally with the first appearance of the genus *Homo,* of which modern humans are the sole descendant.

The Pleistocene epoch witnessed the extinction of several

proboscidean species in Africa, perhaps through environmental change, while the end of the epoch, ten thousand years ago, spelled the end for mammoth and mastodon worldwide, possibly at the hand of man, as we saw in an earlier chapter. That left *Loxodonta* and *Elephas* as the sole representatives of the Proboscidea. Emerging as they did from the evolutionary crucible at about the same time, humans and elephants were destined to have a less than equal relationship, at least from the point of view of the latter.

When we think of the elephants' range in the modern world, we know they live in India and parts of Southeast Asia, and we know of their lands in sub-Saharan Africa. And when we contemplate the plight of these creatures as human populations encroach more and more on their ranges, we often think of it as a problem of the modern world. It's not, as even a brief glance into recent history reveals. Just a few thousand years ago, *Elephas* enjoyed a dominant role in the ecosystems of much of Asia, extending into China and the Middle East. And *Loxodonta* inhabited all of the African continent, not just parts of East, Central, and West Africa, with small populations in the south, as it does now. Just as in the modern world, the twin forces of the inexorable spread of human development and the hunger for ivory were a fatal combination.

Humans have long been fascinated with the beauty—and, probably, supposed magical powers—of ivory. The oldest known carved figure from human prehistory—a tiny horse figurine from Vogelherd in Germany—was hewn from ivory, in this case from the tusk of a mammoth. It was the beginning of an apparently irresistible form of expression, being seen as part of the earliest civilizations in the Middle East, and up through Egypt, Crete, and Greece. But it was the Romans who developed the habit to a high art, not to say excess. They made small figurines with symbolic import; they decorated utilitarian objects, such as combs; they used it as a veneer on large statues and on furniture; they lined rooms with ivory tiles; and they even used it as currency.

The demand for ivory became enormous, with an inevitable result: elephant populations began to shrink. By the second century B.C., the last elephants were killed in northern Africa and in the eastern part of their Asian range. The Romans apparently were unable to make a direct connection between their hunger for ivory and the increasing rarity of the animals that bore it. "Compared to modern societies, the Romans had little information about how wildlife in distant lands was faring," observes the writer Douglas Chadwick in his splendid book, *The Fate of the Elephant.* "While elephants were disappearing due to ivory exploitation, Pliny was writing that the main natural enemies of elephants were dragons."[1] In truth, dragons of a different breed were abroad, breathing a fatal plume of fire in the direction of these large pachyderms.

The Roman Empire, in common with other powerful civilizations of the time, exploited elephants in other ways, too. The animals' strength was an asset in construction and transport, and their power was a cogent weapon in war. However, these uses had little impact on elephant populations compared with the hunt for ivory. And it was small, too, compared with the effect of the destruction of elephant habitat. The ecological face of Eurasia changed dramatically as the need for timber that sustained emerging civilizations fragmented and cleared once extensive woodland and forest. Today's landscape of much of Eurasia and, more recently, the Americas is essentially artificial, the outcome of economic development in forging the modern industrial world. It is well to remember this when offering views on how wildlife habitats in Africa should be maintained.

In the early afternoon of 18 July 1989, Daniel arap Moi, Kenya's head of state, put a flaming torch to twenty-five hundred gasoline-soaked elephant tusks, igniting the fiery destruction of more than $3 million worth of ivory. Some 850 million people around the world witnessed that costly conflagration, either live on television or within days in newspaper and magazine reports. With the heat of the blaze on my back, I conducted a live interview with the ABC television show "Good Morning, America." My message was simple: don't buy ivory, or the elephant will soon be extinct.

I staged this event just four months after being appointed by President Moi to the Wildlife Conservation Department. My immediate goal was to save the elephant, because, by that time, ivory poaching was propelling the African elephant headlong toward extinction. Soon after taking on the post, I faced the decision of what to do with a large cache of ivory that had been confiscated from poachers during the previous four years. Given the desperate state of the finances of the department at the time, the option of selling the ivory on the world market was very attractive. I knew I would be able to do a great deal with $3 million. Equipment was sparse and in a terrible state of repair; fuel supplies were low and in some places nonexistent; anti-poaching units were outgunned by the poachers. Yes, $3 million would have boosted the department's efforts tremendously.

By this time I had seen countless corpses of poachers' victims: great, gray bodies lying bloated under the hot sun, the white excreta of vultures spelling the story of death on their backs. Soon the distended torsos would explode as putrefaction inexorably churned the once-living guts into foul-smelling slime and gas. Worst of all was their faces, transformed from majestic visage into bloody pulp within a few seconds as poachers ripped off their tusks with axes or chain saws. It was a truly sickening and emotion-roiling sight. Add to this the common occurrence of an emaciated, perplexed infant pathetically trying to rouse its murdered mother, and the rational urge to put an end to the slaughter became an emotional obsession. I had to do something, fast, and $3 million would help a lot.

But then I realized, that's not what I needed just then. "No, we won't sell the tusks," I thought to myself one evening while showering. "We'll burn them in a blaze of worldwide publicity." Nothing could be more effective in bringing global attention to the plight of the elephant, I believed. That $3 million would buy global publicity of unmatched value. Debate over what to do about poaching and the ivory trade was at a critical point at this time, and I hoped the spectacle might push it in the direction I considered most likely to be productive. Iain and Oria Douglas-Hamilton encouraged me in my plan, and during one weekend we spent at a friend's ranch in Naivasha we did the simple experiment of putting a piece of tusk on the household fire. At the end of the evening the ivory was black, but not burned. We realized that a much hotter blaze would be necessary if a mountain of tusks was to be set ablaze. A friend took on the job of determining how much gasoline would have to be poured over the tusks so that the blaze would not fizzle out or the president and his attendant ministers be engulfed in a ball of flame. It was a great relief for me that July afternoon that neither of these disasters occurred. It remained to be seen whether the successful blaze would be translated into successful politics and that, in turn, to successful combat on the plains of Africa.

Iain had been "an elephant man" for many years, first studying their behavior in Manyara National Park, Tanzania, and then more broadly throughout Africa, documenting the impact of poaching. His introduction to the latter was in 1969, when he carried out an aerial count of the elephant population in Tsavo National Park, in southeast Kenya. "The dry wilderness was full of red elephants," he wrote. "We flew for hour after hour and there was no end of them. The next day I flew to the eastern border and found a herd of more than a thousand spread out in the bush . . . From here elephants could be found right across the Kilifi district almost as far as the Indian Ocean."[2] The redness Iain referred to was from the rich volcanic soil of the region, which coats the beasts as they roam through classic elephant country, a mix of woodland, plains, and scattered water sources. Iain calculated that there were at least forty thousand elephant in and around the park, making it one of the most important populations in the continent. One reason for Tsavo's pre-eminence as elephant country was the work of its warden, David Sheldrick. A committed man of the wild, he recruited local people as rangers, and they formed a tough, dedicated band of defenders of the mighty species.

Tsavo was not a Garden of Eden, however, tranquil in a balanced, natural harmony. "The park's giant baobab trees, standing out like squat concrete columns, were used as a source of food and water by the elephants during the dry months," recounted Iain. "Now, as the expanding herds opened up the bush, they poked their tusks like swords into the baobab trunks, ripping off the bark to reach the wet pulp within. Many of these magnificent old trees had stood for more than a thousand years. Within a decade, many of them were gone."[3] It wasn't just the baobabs that were suffering; many other woodland trees were being destroyed, too, making parts of the park "look as if a tank force had passed through," as Iain put it.[4] In this situation, Sheldrick faced what all wildlife managers face: How do you *manage* wildlife? Put like this, the question sounds trite, but, as I've hinted earlier, it is not easily answered, given our limited temporal perspective of ecosystems and our limited knowledge of how they work.

Sheldrick strove to understand the scale and nature of the problem. One solution he did not want was that the herd be culled; that is, having bands of hunters shoot a significant proportion of the animals to bring down the population size. Proponents of culling argue that it permits the animals to be maintained in balance with the vegetation on which they depend. Culling was already practiced in some regions of the continent. Against his wishes, Sheldrick had to stand by and watch as a trial culling in Tsavo went ahead, on the orders of his superiors in the department. Three hundred animals were killed. Sheldrick was appalled at what he saw, and managed to put a stop to further carnage. Better to let nature take its course, he decided, even if it meant that many animals would die of starvation from time to time. In any case, he noticed that as the woodlands were opened up by the elephants' activities, other animals, such as zebras and antelopes, thrived in the newly created habitats. Minimal interference was the best way to "manage" the wild, he concluded, a philosophy that I have come to endorse strongly.

The specter of Tsavo heightened debate over the wisdom of culling. Proponents, who saw it as a scientific way to manage wildlife and maintain its habitat, portrayed anticullers as emotional amateurs. Anticullers argued that the cycles of nature should be allowed to take their course. Soon, however, events both natural and human-wrought overshadowed the debate: the first was drought, and the second was intensified poaching.

Just a year after Iain's survey of Tsavo, a terrible drought settled over parts of East Africa. The plains and woodlands were parched, and, denied foliage, starving elephants were driven to eating the trees themselves. The woodlands were ravaged, and soon looked, Iain wrote, "as if they had endured intense shellfire."[5] Emaciated animals roamed the plains in a vain search for food and water, and soon began to die in alarming numbers. At least ten thousand elephants perished in Tsavo alone. Proponents of culling adduced the effects of the drought as a justification for artificially reducing elephant populations. "They would have suffered less in the drought and the habitat would not have been destroyed," said the cullers. "The drought provided a natural means of population control," retorted the anticullers, "and the habitat will bounce back, as it always does in natural cycles."

The anticullers were right, in one respect at least: within a few years new growth was sprouting throughout Tsavo and elsewhere, nurtured by the returned rains. Nevertheless, Tsavo's woodlands were still under pressure, because the elephant population had also bounced back, but for unnatural rather than natural reasons. Fleeing from escalating poaching in the region, elephants sought relative safety within the park's boundaries.

At about this time a landmark meeting of elephant experts took place in Nairobi, organized by Peter Jarman, a research officer in the Kenya Game Department. The gathering was important for several reasons. For one thing, it reported a figure for the elephant population of Kenya, and this represented the first attempt at a national elephant census in Africa. The figure was 167,000. (Within half a dozen years the continentwide population was put at one and a half million.) The second major import of the gathering was a realization of the extent of ivory poaching that was prevalent in Kenya and probably elsewhere. One estimate of ivory export from Africa as a whole in 1973 was $160 million worth, which was the equivalent of 200,000 dead elephants. This was a startling figure, one with stark implications for the future of the species. Mired in a complex political context, this recognition was an extremely sensitive issue for my own country and would become so for other African nations, too.

The bottom line of the meeting was, however, one of conflicting problems. On the one hand, elephants were being killed for their ivory in increasing numbers, and therefore had to be protected. On the other, in many places populations were high enough to cause problems, either damaging habitats, as in Tsavo, or ranging into agricultural areas and destroying crops. Their numbers therefore had to be controlled in these places, it was argued. For a decade and a half these counterposed issues diverted focus from the problem: namely, as long as ivory was available in the marketplace, it would have economic value; and that economic value would be exploited without regard to the fate of the species. With the price of ivory beginning to spiral, the future was clear, or at least it should have been.

. . . .

Between 1973 and 1989 elephant experts argued back and forth over the proper course of action to ensure the survival of the species. It was the conservationists' equivalent of fiddling while Rome burned. Ironically, the international organization responsible for the species' well-being—the Commission on International Trade in Endangered Species, or CITES—was the lead fiddler. One point of contention was the magnitude of the impact of poaching on the continent's elephant population. A second was the feasibility of maintaining a legal trade in ivory while preventing an illegal trade of poached ivory. The whole debate became highly political, which for a while obscured the very real practicalities of conservation that the modern elephant story exemplifies.

The ivory trade operated in a twilight world of secrecy and deceit, with the treasured spoils of slaughtered beasts flowing often clandestinely from Africa to the Far East and on to Europe and the Americas. Whether as trinkets or jewelry, figurines or personal seals, ivory maintained an allure to the human psyche that began millennia ago. Hong Kong was the hub of the trade; there, raw ivory was carved into valued objects. Between 1979 and 1989 more than four thousand metric tons of ivory landed in Hong Kong, a booty that represented perhaps half a million dead elephants. With the price of ivory rising at one point to $120 a pound, the trade netted hundreds of millions of dollars, making some people very wealthy. With this kind of money involved, it is little wonder that many people found ways to circumvent the regulations that were meant to control the trade. It is not my intent to go into this in any detail here. Others have already done that. But it is worth repeating that it did not help matters that the chief agency of control, CITES Secretariat ivory unit, was largely financed by donations from the ivory traders themselves.

When I took over at the Wildlife Conservation Department, debate raged over the future of the trade. Some African nations —South Africa, Zimbabwe, and Botswana—were arguing for its continuation. It would be on a small scale, supplied by ivory from culled animals. CITES was a keen supporter of this position. On the other side of the debate was an increasing number of conservationists who were becoming alarmed at the obvious carnage that was going on at the hands of heavily armed poachers. More

ivory was flowing through the trade route than was legally allow-
able, a sure sign that controls were inadequate in so lucrative a
venture. What was needed were hard data on the impact on elephant populations of the so-called controlled trade. These came
in the form of a continentwide census, an endeavor in which Iain
Douglas-Hamilton played a major role. Iain's data were incorporated in a report by the Ivory Trade Review Group, an independent body funded by half a dozen conservation organizations,
which put the status of the elephant into full public glare in June
1989. Its conclusions were arresting. Let me give some of the
numbers. They speak for themselves.

In the decade up to 1989 the African elephant population
halved, from about 1.3 million to 625,000. This, remember, was
during a time when the ivory trade was supposed to be controlled. The census also demolished the notion—often promoted
by pro-traders—that large populations existed unmolested in
central African forests. These numbers were only a small part of
the story, however. After all, a population of 625,000 individuals
seems numerous and hardly on the brink of extinction. Behind
the numbers was the real impact of the slaughter. The biggest
tusks are sported by mature males, who thus were the prime
target of poachers. Before long, bull elephants were rare. In Kenya, 22 percent of adults were male; in parts of Tanzania, less
than 1 percent. The figure should be 50 percent. The effect on
mating patterns was devastating, as females are fertile for just two
days in their three-month cycle. The chances of a female encoun-
tering a male at the right time diminish with a reduced coterie of
bulls—to virtually zero in the case of the Tanzanian populations.
So, although the figure of 625,000 might look healthy, it por-
tends disaster through the disruption of mating patterns. There's
more.

With big males long dispatched, poachers then turned their
sights on the mature females. The result was obvious in the ware-
houses of ivory traders. In 1979 more than nine hundred metric
tons of ivory were harvested from about forty-five thousand animals; eight years later thirty-five thousand animals died to give
just one third of the quantity, some three hundred metric tons.
Behind the cold statistic of the drop in the average weight of tusk
—from twenty-five pounds to twelve pounds—was the grim reality that fewer and fewer mature females were surviving. As ele-

phant society is matriarchal, with mature females the nexus of group relationships and the repository of experience and knowledge about the world, the loss of such females was shattering. In one Tanzanian population, for instance, only 15 percent of social groups had a matriarch, whereas in undisturbed populations the figure would have been at least 75 percent. Such herds were leaderless and became vulnerable to the vagaries of the environment. That 625,000 figure therefore obscured not only disrupted mating patterns, but also disrupted social patterns. The slaughter of a mature female imperiled the long-term social structure of her group; and it often meant the immediate death of immature offspring that still depended on her for food and protection. All this presented a stark recipe for sudden population collapse.

Some countries bore more of the brunt of poaching than others, of course. In Kenya, the population had plummeted from 167,000 animals in 1973 to fewer than 20,000 in 1989. Barely five thousand remained in Tsavo. Uganda fared even worse, where the ravages of the Idi Amin dictatorship virtually wiped out the entire population. Tanzania lost 75 percent of its population between 1979 and 1989. Against this, countries like South Africa, Botswana, and Zimbabwe witnessed increases in population numbers in this period, a contrast that led to much tension in efforts to ban ivory trade completely. In each of these countries, tight wildlife management often included culling, which, it was argued, not only helped maintain balanced populations and healthy habitats but also provided income to local people through the sale of ivory, skin, and meat. Outside the parks, local people assumed ownership of wildlife in some cases, and were free to exploit it in a limited way. By "exploit" we often mean kill for sport.

In South Africa, where a population of some seventy-five hundred elephants in the Kruger National Park was culled by about six hundred animals a year, income from meat, skin, and ivory was $3 million. The Zimbabwe Wildlife Department netted about $4.5 million by following this practice. Some of the money flowed to the parks, some to the local communities. In all cases, proponents of the system argued, local people were committed to protection of wildlife because they benefited financially from it. This policy is an example of what is called sustainable use, something about which conservationists are becoming more and more heated in their discussions. I am an enthusiastic promoter of the idea that wildlife should be viewed in economic terms. We should use the elephant and wildebeest to generate money for schools, clinics, and agricultural facilities. The question is: How is this best done?

I've long been repelled by hunting of any sort, so the notion of trophy sport is unacceptable to me. And in the case of elephants anyway, so too is culling, the spoils of which bring income in the marketplace. It is impossible to be in the company of elephants for even a short period of time without discerning their sensitivity as individuals and complexity as a group. They are intelligent, social creatures. They have levels and subtleties of communication and understanding that may seem surprising in animals so large. They are not ungainly, lumbering beasts; they are as adept and delicate physically as they are sophisticated in their behavior. For these reasons, I believe it is possible to make an ethical argument against culling elephants. To say no to culling is to recognize this species' rights in a world in which we should coexist.

The nature of that coexistence is, of course, tricky. If you're a farmer who has just lost your entire year's crop to a roaming band of elephants, as sometimes happens, you have a clear notion. "They should not be on my land. Period." As some 40 percent of Kenya's elephants roam on private lands, not parks, this is a not a trivial issue. Westerners should be able to understand this, because few U.S. farmers tolerate bison grazing in their cornfields or wolves roaming among their cattle herds. We know from a study I initiated that of the $420 million that tourism brings annually to Kenya, elephants are responsible for some $30 million. This puts a dollar value on individual animals that is sustainable year after year, not as a one-off bonus through selling their parts. Conservationists and governments face the challenge of using that income both to benefit the community as a whole and to ensure the peaceful coexistence of elephants and villagers. The nature of elephants as large terrestrial herbivores makes that a daunting challenge, because they require a very large range in which to harvest vast quantities of vegetation daily. Biologists know that the smaller a conservation area is, the fewer the species that will survive there. Of terrestrial vertebrates, elephants require among the largest of ranges, so if a park is to provide all they need, keeping them from straying onto private land, it must be very large. Finding and protecting a parcel of wild land that is adequate for the purpose is extremely difficult. But if it can be done, then many other species, with smaller ranges, are protected, too—a conservation piggyback effect.

In October 1989, the debate over the future of the ivory trade culminated in a major gathering of CITES nations, in Lausanne, Switzerland. East African delegates were pressing for an end to the ivory trade; technically this meant elevating elephants to the status of Appendix I, which recognizes that trade in the species critically endangers its existence. Their argument was that controlled trade had been shown a failure; there was too much incentive for illegal trade through poaching and too many loopholes for evading regulations. South Africa, Botswana, and Zimbabwe wanted limited trade based on the products of culling. They argued that if local people are unable to benefit from wildlife, they will lose the incentive to protect it, and the elephant will become extinct. The lead-up to the meeting was tense, as the issue had assumed large dimensions. The elephant had become more than just a species in peril for its survival. It had become emblematic of society's will to conserve nature. It had become, as the biologist David Western termed it, a flagship for conservation. If we are unable to rescue the plain's largest land mammal from the brink of extinction, what chance is there of protecting any species?

While acknowledging the wisdom of having the financial benefits of wildlife flow to local communities, I had no doubt that permitting limited trade would be an open door to further catastrophic poaching, and the elephant would soon be listed not as endangered but as extinct. Many others held the same conviction, but as the meeting began, it was by no means certain which way the vote would go. In the event, the participants at the meeting voted in favor of Appendix I status listing, and a worldwide ban on ivory trading went into effect in January 1990. The price of ivory on the world market plummeted from $120 a pound to a mere $4. I like to think that the fiery signal we sent to the world that July day of 1989 influenced delegates' views for the crucial October vote.

Not so very long ago the enslaving of other human beings was considered by many to be ethically acceptable, but no longer. Not so long ago, the sport killing of chimpanzees and gorillas, our closest evolutionary relatives, was also considered by many to be ethically acceptable, but no longer. I believe that the elephant deserves our respect and protection in the same way.

I have explained my ethical objection to the culling of elephants, and I feel very strongly about this. But there are other objections, too. Proponents of culling frequently speak loftily of the practice as the "scientific" method of wildlife management, offering, they claim, a rational way of maintaining habitats in harmony. In fact, their science is outdated, and derives from the old notion of the balance of nature, a static view of the way ecosystems work. The message I have been developing in this book is that on time scales both long and short, dynamic change, not stasis, is the hallmark of nature. If we are to succeed in our efforts at species conservation and the protection of biological diversity, we need to encompass and exploit that fact. A visit to Amboseli National Park, in southern Kenya, will provide the final punctuation of this point.

A small park by most standards—some 150 square miles in extent—Amboseli is dominated visually by the nearby looming presence of Kilimanjaro, a snow-capped peak rising to 19,340 feet, the highest mountain in Africa. When early European explorers reported their observations back in their home countries, few believed them: "How could a *snow-capped* mountain exist so close to the equator?" Just a few years ago, modern visitors to the region who had studied a map beforehand would have been surprised to find that what is depicted as "Lake Amboseli" was for the most part a dry plain, with low hills dotted here and there. They would not have been disappointed, however, because Amboseli offered a mosaic of habitats that hosted the quintessence of African wildlife. A patchwork of woodland, grassland, swamp, and seasonal streams were home to myriad species of animal. According to Cynthia Moss, who has studied the elephant population there since 1973, these included "elephant, rhinos, hippos, giraffe, buffaloes, zebras, thirteen species of antelopes ranging from the tiny dik-dik to the 1200-pound eland, four kinds of primates, three large cats, wild dogs, two species of hyenas, three species of jackals, several species of smaller cats, plus mongooses, genets, squirrels, hares, and odd species such as the aardwolf, zorilla, ant bear, tree hyrax, honey badger, and porcupine. Over

400 species of birds have been recorded in Amboseli."[6] Countless tourists have photographed these animals, with Kilimanjaro
as a stunning backdrop.

The reference to Lake Amboseli was more historical comment
than reality in 1991. The presence of Kilimanjaro dominates Amboseli not only visually but physically, too, including the enigmatic lake. Five million years ago the region was very different
geologically, not least in the absence of the mountain. Instead, a
large river coursed its way southeast through a relatively flat plain
that is now Amboseli, eventually disgorging into the Indian
Ocean. Sometime between four and two million years ago, volcanic activity heaved the flat plain skyward, eventually building
the modern mountain. As a result, the river was dammed, and a
large lake, some 230 square miles in extent, formed, the recipi-
ent over a period of several million years of countless tons of
sediment and volcanic ash. Eventually, the accumulated debris
filled the lake and diverted the river, beginning a much drier era
for Amboseli. First, an alternation of drying and flooding caused
oscillating habitats in the basin, until some ten thousand years
ago, when desiccation restricted periodic flooding to the western
portion of the lake. This is the basis of the modern habitat.

If Amboseli had to rely on its rainfall for sustenance of its
vegetation, it would be a virtual desert. Kilimanjaro's influence
on the local climate restricts Amboseli's rainfall to about twelve
inches a year, while much more falls on the mountain's slopes.
But what Kilimanjaro takes away by one means it gives back by
another. Underground aquifers bring rain and snow melt from
the mountain, traveling twenty-five miles through porous lava, to
surface cold and clear as life-sustaining springs in the Amboseli
basin. Swamps, pools, and channels sustain a lush vegetation,
ranging from sedge grass to papyrus reed to acacia bush to tall
fever trees. The Amboseli basin's geological history strongly influences its vegetation today, and thus the wildlife that can live
there. The long period of sediment accumulation brought with it
high concentrations of salts. Initially, therefore, only salt-tolerant
plants could survive in the basin once the lake dried up. Eventually, rainfall carried the salts deeper into the earth, allowing less
salt-tolerant vegetation to become part of the ecosystem. However, the salts still lurk there, waiting to rise and fall as the water

table changes, influencing the modern vegetation like deep echoes from the past.

As I describe the Amboseli of modern times, therefore, a visitor would leave with the powerful impression of wildlife teeming on the plains and in the woodland, against the backdrop of the highest mountain in Africa. Five million years ago, it would have been very different: A tree-lined river coursing across a flood plain, with a different menagerie of animals from those of today, and no mountain. Even ten thousand years ago it would have been dramatically different, with a major lake shimmering at the foot of the great mountain. Vegetation and animal life would have been rich, but the vegetation would have been limited to salt-tolerant plants, restricting the diet available to the animals there. It may seem trite to emphasize the great changes over long periods of time, but I emphasize them because it is so easy to be awed by the majesty of modern geology and forget that it, like us, is ephemeral in the scope of evolutionary time. In a few million years Mount Kilimanjaro will be a mere remnant of itself, the victim of erosion, time, and change.

I want to make two further points about Amboseli as specific illustrations of general patterns relating to conservation and biodiversity. The first is that even within short periods of time, dramatic changes in habitat can occur. I'm not talking about the appearance of a mountain or the disappearance of a lake; I'm talking about something more modest, often linked to rainfall patterns. Nevertheless, even these modest changes can have an enormous impact on the kind of ecosystem that survives. The second is that the rich mosaic of habitats I described earlier, populated as they were with a high diversity of animal species, is not simply a product of an interaction of the terrain and water supply. Something else was instrumental in shaping the habitat and building a high diversity there: *Loxodonta africana.*

When I initially described the modern Amboseli I said that, just a few years ago, visitors to the region would have been surprised at the disparity between what the map promised and the reality. During the preceding decade the region had changed startlingly, becoming much drier and with a fraction of the vegetation cover. The swamps were lower, the papyrus was thinned out, and, most stark of all, the woodlands were devastated. In the central part of the park, half the species of plant that had thrived there a decade earlier were gone. Dust devils skittered among skeletal tree trunks. Several species of bird, some small mammals, and some large herbivores had gone, too. Gazelles, wildebeest, and zebras—all grazers, dependent on grass—still were plentiful, but browsers, such as giraffe, impala, and kudu, were uncommon. Vervet monkeys, once common, were rare, and the population of baboons had more than halved. Amboseli was no longer an African Eden. Almost certainly, several agencies contributed to the devastating change, although there is disagreement over what they were. One commonly fingered culprit was the elephant, caught between poaching pressure outside the park and reduced vegetation within. Not surprisingly, the hoary issue of culling the population arose again.

It is true that the elephant population in the park was high, possibly the highest density anywhere in the continent: some 750 individuals in the 150 square miles. It is also true that much of this population increase was the result of animals fleeing the threat of poaching beyond the park's boundaries. The animals were safe in the park, and they knew it. Inevitably, such a high density of elephants imposed pressure on the vegetation: after all, an individual requires 375 pounds a day to sustain it in good health. That pressure is exacerbated if the water supply is diminished, as it appeared to be. Nevertheless, I suspected that it was too simplistic to blame Amboseli's high elephant population completely for the habitat devastation there. There was good reason to believe that what we were seeing was part of the cycling that goes on in nature. For instance, when the explorer Joseph Thomson traveled through the region a century ago, Amboseli was much more as it is in 1991 than in 1981. In his book *Through Masailand,* which was published in 1887, he described the terrain as desolate and barren. It was, he said, flat and dusty, with no trees, a description that could have been applied a few years ago. The game was plentiful, however, although almost certainly less so than during more lush times.

Even if we knew nothing more about Amboseli than these descriptions from the 1880s, the 1980s, and the early 1990s, we would have to conclude that some kind of natural cycling was going on: from dust bowl to lush mosaic habitats to dust bowl again. As it is, we have sufficient historical information to know that during the last century the vegetation of the park fluctuated dramatically, shifting from dense woodland to barren plains, as it did elsewhere in East Africa. Such fluctuations are usually a response to rainfall patterns. But in the case of Amboseli, changes in the water table occasionally brought those ancient salts close to the surface, where they were inimical to tree roots, causing a slow death. I suspect that such a shift in salt distribution below ground contributed to what we have recently seen above the surface of the ground. I'm not discounting the impact of a large elephant population on Amboseli's habitat, and also the pressure of hundreds of thousands of tourists that goes with it. But I would argue that it is a contributing factor, not the whole cause. If this is the case, then short-term efforts to ameliorate the situation—through culling, for instance—are likely to be futile. A light hand —and much patience—is much the wiser means to deal with wildlife management, both in times of stress and in more equable periods.

I developed this philosophy early on in my tenure as director of the Kenya Wildlife Service. Amboseli and its history seemed to speak clearly of nature's ever-changing face, a warning to those who would try to bend nature to their will. And, as if to emphasize the point, Amboseli has changed again. In the last three years the water table has risen once more, the swamps are expanding, lakes sparkle where dust devils used to spiral through the parched earth, and trees and bushes are sprouting bountifully. There is less pressure from elephants, too, because the cessation of poaching has allowed the animals to wander safely beyond the park's boundaries. And some of the species that disappeared from the park are returning. This land at the foot of Kilimanjaro offers a lesson we must heed if we are to understand the flow of life of which we are a part.

When you watch an elephant dismember a tree or uproot a seedling, it is easy to be impressed by the animal's destructive power. Certainly, the scenes I described of Amboseli a few years ago speak of mass destruction. But biologists have now come to view elephants' destructive power in a different way; that is, as a creative power. The idea is as simple as it is powerful. The effect of destroying trees in woodlands is to create an environment in which bushes can survive; the destruction of bushes in a bushland savannah opens up space for grassland to sprout. Anyone who has visited a dense forest cannot fail to be impressed by its majesty, as the tall trees loom above you, perhaps drooping with epiphytes. But after a while the odd stillness of it all begins to penetrate the consciousness. Where are the large animals? They aren't there, or at least not very many kinds of them; not the grazers and browsers you see in open woodland and grassland.

Elephants, biologists now realize, are a key agency in the creation of mosaic habitats in which other species can thrive: the browsers in the newly created bushland, the grazers in newly created grassland. Biologists use the term *keystone herbivore* to describe this kind of species. And, just as an arch will collapse if the keystone is removed, so too will an ecosystem if its keystone herbivore goes extinct. Norman Owen-Smith, a biologist at the University of the Witwatersrand in Johannesburg, claims it is possible to see this on a grand scale in the not very distant past. Almost certainly, human predation was responsible for the extinction of the large herbivores in the Americas at the end of the Pleistocene, ten thousand years ago, concedes Owen-Smith. But, he points out, many smaller mammals and birds became extinct too, species that would have been unlikely victims of hunters preoccupied with mastodon and gomphotheres. With the megaherbivores eliminated, open forest glades closed up, bushland reverted to woodland, and grassland mosaics became uniform tall grassland. These vegetational changes, which can be discerned in the fossil pollen record, would have restricted the habitats available to smaller herbivores, causing the cascade of extinctions we see in the fossil record.

It is possible today to see similar patterns locally where elephants have been eliminated in Africa. In the absence of elephants, rich habitats populated with browsers and grazers become uniform bushland or woodland, supporting far fewer species. This is apparent north of Amboseli, for instance, where the terrain is uniform thick bush. The implication is clear. If through horrendous lack of foresight or lack of will, the elephant becomes extinct, many more species will follow its path to evolutionary oblivion. The elephants' effect as a keystone herbivore has been overlooked in the high-decibel debate over the future of the ivory trade, the wisdom of culling, and so on, and yet it is a vital part of the equation. The elephant is not just a single species in danger of extinction. It is not even just a flagship of conservation, a test of our will to respect and preserve nature. Upon its survival rests the survival of many other species that constitute the great diversity of African wildlife that is our heritage of millions of years of evolution.

Nothing more, nothing less.

The Future

A SYNTHESIS of the new insights in evolutionary biology, ecology, and paleontology reveals the true nature of *Homo sapiens* in the world: we are a mere accident of history.

Homo sapiens has become the most dominant species on Earth. Unfortunately, our impact is devastating, and if we continue to destroy the environment as we do today, half the world's species will become extinct early in the next century.

Even though *Homo sapiens* is destined for extinction, just like other species in history, we have an ethical imperative to protect nature's diversity, not destroy it.

12

An Accident of History

W E ARE IN THE MIDST of a seismic shift in thinking about the nature of ourselves and the world we live in. It is no hyperbole to describe the magnitude of the shift as an intellectual revolution. In earlier chapters I assembled the components of that shift, so now we can begin to see more clearly the nascent picture as a whole. Separate, shadowy lines are coalescing into a unified, sharp focus, and we can see that the image portrays a concept of life's flow that is radically different from the view that has held sway for so long. Be patient, as a little reiteration of earlier themes in the book is necessary if we are to illuminate this new concept as it deserves.

We need courage if we are to embrace this new concept—in reality, a series of concepts—because it requires that we discard many comforting notions about our place in the universe of things. The payoff of perceiving ourselves clearly within the new intellectual framework is enormous, however, not only because we are moving toward a fuller understanding of the world, which has been an engine of human endeavor for millennia; but also because it equips us more efficiently for steering humankind into the next century and beyond. It is this future—the *immediate* future in the grand scheme of things—that this final section addresses. I will state boldly right now that I believe we face a crisis —one of our own making—and if we fail to negotiate it with vision, we will lay a curse of unimaginable magnitude on future generations.

This chapter will formulate the nature of the intellectual revolution that swirls around us now, and its implications for a deeper insight into the flow of life, which we ride in company with many fellow passengers. The final two chapters will reveal why we are at a watershed in human history; indeed, how we have constructed that watershed, with one slope leading to a catastrophe unprecedented in Earth history, and the other to a degree of harmony with the world of nature. If I sound alarmist, it is intentional. I will also develop further what is unique about this book: that is, contemplating *Homo sapiens* in the context of Earth history and, by extrapolation, Earth's future. It is not an easy exercise, because the human mind is used to thinking in terms of decades or perhaps generations, not the hundreds of millions of years that is the time frame for life on Earth. Coming to grips with humanity in this context reveals at once our significance in Earth history, and our insignificance. There is a certainty about the future of humanity that cheats our mind's comprehension: one day our species will be no more.

For much of this century the twin sciences of evolutionary biology and ecology have been imbued with a view of life that, quite frankly, offered reassurance to the human psyche. That view derived from the revolutionary and powerful thinking of Charles Darwin and Charles Lyell, great naturalist and great geologist, respectively. Earth history, in both the biological and geological realms, progressed smoothly and gradually through immense tracts of time, they said. The steady accumulation of tiny changes builds all the vast geological formations in our world, such as deep canyons and high mountains; and they populate the continental surface and the deep sea with an extravagant diversity of life, of which *Homo sapiens* is but one species among many millions. In order to understand the strength of Darwin and Lyell's conviction of their gradualistic worldview, it is important to remember that they were reacting to an earlier intellectual tradition, one that held the Earth and its inhabitants to be the hapless products of occasional crises, or catastrophes. As each catastrophe visited, gargantuan floods or convulsions of the continents consigned Earth's biota to oblivion, to be replaced by a coterie of new, more advanced forms of life.

Although not all proponents of catastrophism framed their ideas in a religious context, the hypothesis inevitably was so imbued: divine forces wrought destruction just as they fueled creation. By no other means could the Earth and the living creatures therein be formed, it was widely believed. When Lyell showed that even the most gigantic of geological structures could arise slowly over long periods of time, and Darwin formulated a theory for the origin of myriad forms of life, slowly over long periods of time, they jointly offered a perspective—gradualism— that was apparently more scientific than catastrophism, as it did not invoke divine intervention. Catastrophism was dead. The age of modern biology and modern geology was born, with the all-pervading concept of gradualism its midwife.

Most important of all, and central to the Darwinian view, is that life's flow is guided by an incessant struggle for existence, in which the fittest thrive and the less fit succumb. Species are engaged in a constant battle, both with other species and with their physical environment, twin challenges to which natural selection responds by molding appropriate adaptations. Survival is a mark of success in the struggle, extinction a mark of failure. I cited many of Darwin's statements on this point in earlier chapters, but I'll repeat one of them here, because it encapsulates a sentiment that was clearly important when it was written, and continued to be so for more than a century: "As natural selection works solely by and for the good of each being, all corporeal and mental endowments will tend to progress towards perfection."[1]

Two important concepts flow from this view. The first is that the history of life that we see written in the fossils in the ground is a record of that struggle, with a steady succession of losers and winners in life's race, from the earliest to the most recent of times. Because natural selection builds adaptations by tiny increments, the trajectory of evolutionary change is slow and steady. The second concept is that, because life's flow is steered by the success of ever more superior forms, the direction and shape of the flow is in an important sense inevitable. The progression of fish, to amphibians and reptiles, to mammals—and eventually to us—reflects an unfolding of life not just as it was but as it must have been. In this view, the element of chance, or randomness, plays no part in the direction and shape of life's flow, and this includes the eventual appearance of *Homo sapiens*.

I suggested earlier that this traditional view was comforting to the human psyche, and we can see why. We humans dislike uncertainty and are repelled by randomness, especially in relation to our existence. We hate to think that our being is the outcome of the throw of evolutionary dice. The traditional view I just described offers us solace, because it says, first, that our very existence is testimony to our superiority; and, second, that life was leading in our direction all along. Remember how the paleontologist Robert Broom put it in his 1933 book, *The Coming of Man: Was It Accident or Design?*: "Much of evolution looks as if it had been planned to result in man, and in other animals and plants to make the world a suitable place for him to dwell in."[2] Broom's was an extreme statement, but it captures the essence of a mode of thinking that pervaded all scholarly endeavor.

We now have to face the fact that this is wrong. There was no steady progression from simple to complex forms throughout Earth history. Simple forms of life arose early on, it's true. But, as we saw in chapter 2, that early simplicity continued in mind-numbing sameness for billions of years, with nothing more complex than single-celled organisms for six sevenths of Earth history. When complexity eventually arose 530 million years ago, in the form of multicellular organisms, it did so explosively; within five million years (an instant in geological time), evolutionary innovation produced a myriad of multicellular forms of life. Life's flow is therefore not smooth, but extremely erratic. (In fact, as Stephen Jay Gould and Niles Eldredge argued in their theory of punctuated equilibrium, many and perhaps most evolutionary innovations occur in rapid bursts of change rather than gradually, over long periods of time.)

The reality of the Cambrian explosion was only mildly disturbing to our psyche, however, because we could cling to the notion that, erratic though it may be, the direction of life's flow was still in a sense predictable and we were still its inevitable product. The new insights into the history of life following the

Cambrian explosion, as we saw in chapter 3, however, deprived us of this line of argument. The wild experimentation of the Cambrian explosion yielded as many as a hundred different forms of life, or body plans. Within a few million years, only a fraction of those forms remained, to become the motes in the kaleidoscope of life today. Had the survivors avoided extinction through inherent superiority, we could comfort ourselves with the knowledge that we are the descendants of deserving winners. But we are not. There was nothing obviously superior in the survivors or inferior in the victims of that early mass extinction. It was, as Gould put it recently, "the largest lottery ever played out on our planet,"[3] and we happen to be descendants of one of the lucky winners. We share today's world with descendants of other lucky winners. Play the lottery again, and a different coterie of winners would emerge, yielding a different set of body plans upon which modern life would be based. No doubt many would be outlandish in the extreme, to judge from the odd life forms that vanished during that early extinction. We therefore have to accept the fact that the living world of which we are a part is but one of countless possible worlds, not the only inevitable one. It is, rather, simply a contingent fact of history.

One of the most important components of the current intellectual revolution is a new understanding of the nature of extinction. Darwin, remember, was suspicious of talk of mass extinctions, and he explained the apparent evidence for mass dyings in the fossil record as an artifact of an incomplete record. Mass extinction smacked of catastrophism, and that was anathema to him. Eventually, the reality of occasional mass extinctions became irrefutable, denting further the cherished idea of a smooth passage to life's flow. As we saw in chapter 4, five major biotic crises punctuated that flow, hurling as many as 95 percent of existing species into extinction in a geological eyeblink. Oddly, even though mass extinctions became recognized as a major factor in the patterns of the history of life, their nature remained a neglected field of study. They were considered to be complex and difficult to understand. Nevertheless, biologists assumed that, whatever their cause, these events intensified the struggle for existence, consigning unusually large numbers of species to evolutionary oblivion within a short period of time. Mass extinction was held to be just like normal, or background, extinction, but on a larger scale. And, just as in normal times, survival through such events was deemed a mark of adaptive superiority, extinction a mark of adaptive inferiority.

The recognition, just in recent years, that mass extinctions do not represent the processes of background extinction writ large must rank as one of the more important discoveries in evolutionary biology of this century. Whatever their cause, mass extinctions operate by different rules from those prevailing during background extinction. Darwinian evolution, important in background times, is suspended during biotic crises. Survival through these events depends not on the quality of a species' adaptation, but on such properties as the geographical distribution of groups of species, or clades (localized clades are vulnerable; widespread clades fare better, no matter how many species they contain) and body size (large species are more vulnerable than small ones). The inescapable conclusion is that during mass extinctions, a species' survival has as much to do with good luck as good genes, to use David Raup's catchy phrase.

As I indicated earlier, the mammals outlived the dinosaurs, not through inherent superiority, but because that was the roll of the dice. Their small size no doubt contributed to their survival. Not all mammalian lineages made it through that event sixty-five million years ago, of course. Some went extinct. And the fact that the primate lineage was among the survivors was, again, a matter of luck, not superiority. Had *Purgatorius* been among the losers, there would have been no prosimians, no monkeys, no apes— and no humans.

There is no escaping the conclusion that, special though *Homo sapiens* may be in many ways—particularly in our creativity and our consciousness—our existence was not inevitable. Worse, it involves a degree of randomness that many will find difficult to accept. But it's true. As I put it in chapter 5, "We are but one of millions of species here on Earth, product of half a billion years of life's flow, lucky survivors of at least twenty biotic crises, including the catastrophic Big Five." From a parochial point of view, this is one of the more profound insights of the current intellectual revolution, but, for the living world as a whole, it is not the most important.

Mass extinctions were once viewed as major interruptions in the flow of life, dealing a death blow to many species, but not shaping evolution. Biologists believed that natural selection does the shaping. But we know now that the rules of death are different during these times, and survival involves an important element of randomness. These factors, not natural selection, determine which species survive and which don't, making mass extinctions a major force in shaping patterns in the history of life. Indeed, among the agents influencing the overall pattern, they are *the* force. As David Raup observed, "Extinction, and especially mass extinction, can be seen as a vital ingredient in the evolution of complex life as we know it."[4] Of the components that constitute the current intellectual revolution, this realization is extremely important. But there is more.

Several times I've used the phase "whatever their cause" in relation to mass extinctions. This qualification is a recognition of the fact that the genesis of these events is indeed extremely complex and difficult to understand, and is currently one of the most hotly debated topics among evolutionary biologists and paleontologists. I described earlier some of the agencies that may be involved, such as global temperature change and regression of sea level. But, as we saw in chapter 4, there is now no denying that at least one event—the Cretaceous extinction, sixty-five million years ago—and perhaps several (or, in the extreme, all) was or were triggered by Earth's impact with a giant asteroid or comet. The recognition that mass extinctions play so vital a role in shaping Earth history was important enough in the development of evolutionary theory. The notion that these events may be the result of extraterrestrial impact is truly compelling. We are forced to leave behind a Darwinian world that is shaped by forces we can understand and identify with in our daily existence, and accept one that is the hapless victim of a fickle universe. Gone is an image of the flow of life as smooth and predictable, with humans its inevitable culmination; its replacement is a world that is erratic and unpredictable, and in which our place is achieved through a large slice of luck. Catastrophism is back with us, and it is real.

I spoke of the twin sciences of evolutionary biology and ecology in the context of the current intellectual revolution, but so far have addressed only the former. A similar shift in worldview has occurred in ecology, particularly in relation to how ecological communities come to be the way they are. In chapter 9 I said

that, until recently, ecologists believed that ecological communities are the way they are because that's the way they should be. I
acknowledge that this is putting the position rather simplistically,
but what I mean is that it is analogous to the Darwinian view of
the inevitability of life's flow through evolutionary time. In the
briefer time frame of the assembly of ecological communities,
the rules that were held to operate predominantly were the same
as those which produce adaptation through natural selection:
species interact with each other and with the physical environment, producing a harmony that was encapsulated in the phrase
"the balance of nature." Ecologists believed that the makeup of
a community in a particular location reflects the conditions of
that location, and it is different from another community in a
different location with different conditions.

It is obviously true that the interaction of species within a community is important in its composition: fungi sustain plant roots;
plant leaves sustain insects; insects sustain birds; and so on. And
it is also true that species are adapted to certain local physical
conditions. But, as I described in chapter 9, ecologists now recognize that these influences are only part of the explanation of why
communities are the way they are. Just as evolutionary biologists
have had to accept that randomness plays a significant role in
life's flow, so, too, have ecologists in their own realm of interest.
The members of an ecological community are not nicely poised
at equilibrium with one another, in a balance of nature. Much of
the shape and behavior of a community is determined by chaotic
interaction and by the emergence from within the community of
properties—such as the resistance to invasion—that defy ready
explanation. The former view is of communities that are predictable but static. The latter is of unpredictable—even mysterious—
communities, and they are dynamic. And this dynamic state con-
tributes to the biological diversity of the world, which, ultimately,
is our concern here. Counterintuitively, constant change—the
dynamic state—is the source of long-term stability in communities; try to block change in the short term, and you ensure inimical change in the long term.

Humans long for predictability, in relation to the world of
nature around us and, most particularly, in relation to our own
existence and our future. But it is obvious that, in the realm of
evolutionary biology and ecology, ours is an unpredictable world

and our place in it an accident of history; it is a place of many possibilities that are influenced by forces beyond our control and, in some cases at least, beyond our immediate comprehension.

Ours is a less certain world than we thought it was, but it is also more interesting for that.

13

The Sixth Extinction

A N ACCIDENT OF HISTORY we may be, but there is no question that *Homo sapiens* is the single most dominant species on Earth today. We arrived late on the evolutionary scene and at a time when the diversity of life on the planet was near its all-time high. And, as we saw in chapter 10, we arrived equipped with the capacity to devastate that diversity wherever human populations traveled. Blessed with reason and insight, we move toward the twenty-first century in a world of our own creation, an essentially artificial world in which (for some, at least) technology brings material comfort and leisure brings unprecedented artistic creation. So far, unfortunately, our reason and insight have not prevented us from collectively exploiting Earth's resources—biological and physical—in unprecedented ways.

Homo sapiens is not the first living creature to have a dramatic impact on Earth's biota, of course. The advent of photosynthetic microorganisms some three billion years ago began to transform the atmosphere from one of low oxygen content to one of relatively high levels, reaching close to modern levels within the last billion years. With the change, very different life forms were possible, including multicellular organisms, and previously abundant forms that thrived in a low oxygen environment were consigned to marginal habitats of the Earth. But that change was wrought not by a single, sentient species consciously pursuing its own material goals, but by countless, non-sentient species, collectively and unconsciously operating new metabolic pathways. The reason and insight that emerged during our evolutionary history bestowed a behavioral flexibility on our species that allows us to multiply bounteously in virtually every environment on Earth. The evolution of human intelligence therefore opened a vast potential for population expansion and growth, so that collectively the almost six billion humans alive today represent the greatest proportion of protoplasm on our planet.

We suck our sustenance from the rest of nature in a way never before seen in the world, reducing its bounty as ours grows. We are, as Edward Wilson has put it, "an environmental abnormality." Abnormalities cannot persist forever; they eventually disappear. "It is possible that intelligence in the wrong kind of species was foreordained to be a fatal combination for the biosphere," ventures Wilson. "Perhaps a law of evolution is that intelligence usually extinguishes itself."[1] If not a "law," then perhaps a common consequence. Our concern is: Can such a fate be avoided?

When I talk about reducing nature's bounty, I'm referring to the extinction of species that is currently occurring as a result of human activities of various kinds. In chapter 10 I described the trail of biotic destruction humans left in their wake as they swept into new environments in the prehistoric and historic past: settlers of new lands extirpated huge numbers of species, through hunting and clearing of habitats. Some modern scholars argue that this was but a passing episode in the human career and that, despite massive population expansion today, talk of continued species extinction is fallacious. It should be obvious from the tone of the preceding few paragraphs that I am not among their number. I believe that human-driven extinction is continuing today, and accelerating to alarming levels.

In the remainder of the chapter I will develop the argument for my concern. In the final chapter I will ask whether or not it matters to us and our children that as much as 50 percent of the Earth's species may disappear by the end of the next century. I will also address the longer-term future, which puts our species in a larger geological context with the rest of the world's inhabitants. And I will suggest that the insights we have gained from the current intellectual revolution I formulated in the previous chapter demand that we adopt a certain ethical position on the impact of *Homo sapiens* on the biodiversity of which we are a part.

Humans endanger the existence of species in three principal ways. The first is through direct exploitation, such as hunting. From butterflies, to song birds, to elephants, the human appetite for collecting or eating parts of wild creatures puts many species at risk of extinction. Second is the biological havoc that is occasionally wreaked following the introduction of alien species to new ecosystems, whether deliberately or accidentally. I talked earlier about the biological convulsion experienced by the Hawaiian archipelago through countless species of birds and plants taken there by the early Polynesians and later by European settlers. A devastation of equal magnitude is currently under way in Africa's Lake Victoria, where more than two-hundred species of fish have disappeared within the past decade. The Boston University ecologist Les Kaufman, who has studied the event in great detail, calls it "the Hiroshima of the biological apocalypse, the demonstration, the warning that more is on the way."[2] Several interacting factors are involved, such as overfishing and pollution, but the major culprit is the voracious Nile perch, which was introduced to the lake for commercial fishing some four decades ago.

   The third, and by far the most important, mode of human-driven extinction is the destruction and fragmentation of habitat, especially the inexorable cutting of tropical rainforests. The forests, which cover just 7 percent of the world's land surface, are a cauldron of evolutionary innovation and are home to half of the world's species. The continued growth of human populations in all parts of the world daily encroaches on wild habitats, whether through the expansion of agricultural land, the building of towns and cities, or the transport infrastructure that joins them. As the habitats shrink, so too does the Earth's capacity to sustain its biological heritage.

   The Oxford University ecologist Norman Myers was the first to call wide attention to the impending catastrophe of deforestation, in his 1979 book, *The Sinking Ark*. If the rate of tree felling continued at its prevailing rate, which Myers estimated to be as much as 2 percent a year, the world would "lose one-quarter of all species by the year 2000," he wrote. A further century would add a third of the remaining species to the death toll. The decade and a half since *The Sinking Ark's* publication has witnessed roiling debate over the reality of the numbers. Are the forests disappearing at the rate claimed? Even if they are, would 50 percent of the world's species really disappear?

Initially, Myers's (and others') prognostications received a sympathetic hearing, and eventually built a sense of genuine alarm and concern among biologists and politicians. Grave statements flowed from weighty bodies. "The species extinction crises is a threat to civilization second only to the threat of thermonuclear war," warned the Club of Earth in a publication released at the beginning of a major conference of biodiversity, held in Washington, D.C., in September 1986. A recent joint statement by the U.S. National Academy of Sciences and the Royal Society of London must qualify as the most prestigious: "The overall pace of environmental change has unquestionably been accelerated by the recent expansion of the human population . . . The future of our planet is in the balance." Individual ecologists were equally emphatic. I'll quote two of the most prominent. Stanford University biologist Paul Ehrlich said at the Washington conference, "There's no controversy among mainstream biologists that there is a crisis in biodiversity." At that same gathering, Edward Wilson stated that "virtually all students of the extinction process agree that biological diversity is in the midst of its sixth great crisis, this time precipitated entirely by man."

Just recently, however, a backlash has developed, with the doomsayers being accused of overstating their case or, worse, fabricating it. Articles have appeared in several periodicals, expressing skepticism of the alleged danger. An article titled "Extinction: are ecologists crying wolf?"[3] was recently published in *Science,* for instance; and the 13 December 1993 issue of *U.S. News and World Report* ran a cover story, titled "The Doomsday Myths." These and other articles essentially suggest that although ecologists *believe* that many species are becoming extinct, or are about to become so, they don't actually *know* for sure. Julian Simon, at the University of Maryland, has been saying as much for a decade, and his voice has become even louder of late. The most prominent of the anti-alarmists, Simon wrote in a 1986 article, "The available facts . . . are not consistent with the level of concern."[4] In a debate with Norman Myers in New York in 1992, Simon repeated this view: "The actual data on the observed rates of speciation are wildly at variance with . . . the purported danger."[5] He was more direct in an opinion article he published in the 13 May 1993 issue of the *New York Times:* he described claims by various ecologists that current extinction rates were equivalent to those of a mass extinction as "utterly without scientific underpinning" and "pure guesswork."[6] Professor Simon is the Dr. Pangloss of the environment.

Why has there been this criticism of scientists whose expertise supposedly is the understanding of the dynamics of biodiversity? Perhaps one reason is that the message is so startling that people are simply unwilling to hear it, or, if they hear it, are unwilling to believe it. A human-caused mass extinction *is* startling. Ecologists' predictions therefore came to be viewed as "the outpouring of overwrought biological Cassandras," says Thomas Lovejoy, of the Smithsonian Institution.[7] Another reason for the incredulity, no doubt, was the disparity of predictions from different authorities of the scale of the imminent extinction, which ranged from 17,000 species lost a year to more than 100,000. If the experts are so uncertain about the magnitude of the alleged extinction, critics legitimately wondered, how can we believe *anything* they say? I'll come back to this.

There is, I suggest, a further reason, one having to do with uncertainty of a different nature: that is, about ourselves. If we accept that species can be pushed into extinction as easily as the ecologists are telling us, then perhaps the tenure of *Homo sapiens* is less secure than we would like to believe. Perhaps we, too, are destined for extinction. We dislike uncertainty about our origins; and we dislike uncertainty about our future even more.

The two pertinent questions, remember, are these: Are the tropical forests being felled at a rate near to what Norman Myers and others claim? If so, what is the impact on the species living there? The first is the easier of the two to answer directly, principally because it can be observed directly.

Myers's 1979 estimate of 2 percent of standing forest being cut each year was based on a compilation of piecemeal observations in various parts of the world, and extrapolation from these to the rest of the world. This proportion works out to be some eighty-thousand square miles a year, or more than an acre a second. Dozens of studies carried out during the 1980s and early 1990s attempted to test this contention. Some claimed it to be an over-estimate, some an underestimate. Now, with the use of extensive satellite imagery of much of the world's land surface, the answer is beyond reasonable doubt. For instance, two independent reports in the early 1990s, one by the World Resources Institute, Washington, and the second by the United Nations Food and Agriculture Organization, each produced figures in the range of eighty-thousand square miles of forest lost each year. (This is 40 to 50 percent higher than a decade earlier.) At this rate of destruction, tropical forests will be reduced to 10 percent of their original cover soon after the turn of the century and to a tiny remnant by 2050. Only a deliberate obscurantist would deny these numbers.

A reduction of this magnitude is bad enough for the survival of species in the forests, but there is worse news. A more recent satellite study reveals that even where forest is not clear-cut, it is often fragmented into small "islands" that are ecologically fragile. In an epic experiment begun in the late 1970s in the Brazilian forest, Thomas Lovejoy and his colleagues have been studying the ability of such islands of different sizes to sustain species. With islands varying in size from 2.5 acres to 25,000 acres, the venture is the biggest biological experiment in history. One of the expected observations is that species would become extinct more rapidly and more extensively in small patches than in larger ones. Some of the vulnerable species are those which require a large range, for various reasons. And, as we saw in earlier chapters, extinction of these species often causes other species to become extinct, too, even though they themselves don't require large territories. For instance, three species of frog vanished from one 250-acre plot early in the experiment, because the habitat was too small to support peccaries, whose wallowing in mud created ponds for the frogs. Such cascades of extinction continue for many years after the island plot is established. Other species may be vulnerable to extinction in small islands, because of the small population sizes that can be sustained there. Small populations can fall victim to sudden bouts of disease or external perturbations, such as storms, whereas large populations can weather such events.

An unexpected finding from the experiment, however, is that even large forest patches are less sturdy than might be imagined. The reason is the so-called edge effect. Habitats deep in the forest enjoy a degree of protection from external perturbation, whereas those at the boundary between forest and grassland, for instance, are exposed to winds, dramatically varying microclimates over short distances, incursion by nonforest animals and human hunters, and other inimical circumstances. The result: species of animals and plants are vulnerable to extinction for as much as a half a mile into the forest. The edge effect is therefore important even for large tracts of forest. This discovery has become especially important with the new satellite survey, which shows that logging has been leaving a vastly greater proportion of Amazonian tropical forests vulnerable to edge effects than was realized. "Implications for biological diversity are not encouraging and provide added impetus for the minimization of tropical deforestation," the investigators reported in *Science*.[8]

The key variable in the equation, then, is the effect of forest loss and fragmentation on species survival. Before I go into this, however, it is important to emphasize that habitat loss is not confined to tropical forests. For instance, a study by the U.S. National Biological Service reported in February 1995 that during this century half the country's natural ecosystems had been degraded to the point of endangerment. Entire communities are now on the brink of extinction. In a second study, published a few months later, the service noted that "if unchecked, human activities will continue to result in an upset balance of species interactions, alterations of ecosystems and extensive habitat loss." Evidently, concern for the future of our biological heritage has to be played out in all countries of the world, not merely in the poorer, developing countries.

As I said earlier, the growth of human population worldwide is encroaching on wild habitat, both for constructing villages, towns, and cities, and the infrastructure that goes with them, and for producing food, both plants and livestock. Human population has expanded dramatically in recent history, as everyone is aware. From half a billion in 1600 to a billion in 1800; by 1940 it had reached almost 3 billion; in the past fifty years it doubled, to 5.7 billion; and it is set to double again in the next half century, to more than 10 billion. If all these people are to enjoy a standard of living above the poverty level that prevails in many of the less developed regions of the world today, the global economic activity will have to rise at least tenfold. At what cost?

Even today, humans consume 40 percent of net primary productivity (NPP) on land; that is, the total energy trapped in photosynthesis worldwide, minus that required by the plants themselves for their survival. In other words, of all the energy available to sustain all the species on Earth, *Homo sapiens* takes almost half. To the Stanford biologists Paul and Anne Ehrlich, the implications are ominous. "What a substantial expansion of both the population *and* its mobilization of resources implies for the redirection and further loss of terrestrial NPP by humanity is obvious," say the Ehrlichs. "People will try to take over all of it and lose more in the process."[9] For every extra 1 percent of global NPP commandeered by our species in the coming decades, a further 1 percent will become unavailable to the rest of nature. Eventually, primary productivity will fall, as space for the producers falls, and a downward spiral will eventually kick in. The world's biological diversity will plummet, including the productivity on which human survival depends. The future of human civilization therefore becomes threatened.

Not everyone accepts this doomsday outlook, of course, most particularly Julian Simon. In what must rank as one of the more daring and optimistic predictions ever made, Simon declared the following in the debate with Myers: "We now have in our hands . . . the technology to feed, clothe, and supply energy to an ever growing population for the next 7 billion years."[10]

One of these scenarios—the imminent threat of doom or essentially infinite human expansion—must be wrong.

The method by which ecologists calculate the fate of species in habitats that are reduced in size is based on island biogeography theory, which the Harvard biologists Robert MacArthur and Edward Wilson developed in 1963. Partly the outcome of empirical observation, partly mathematical treatment, the theory is the foundation of much of modern ecological thinking. "We had noticed that the faunas and floras of islands around the world show a consistent relation between the area of the islands and the number of species living on them," Wilson recalled recently. "The larger the area, the more the species."[11] MacArthur and Wilson saw this relationship wherever they looked, from the British Isles to the Galápagos Islands to the archipelago of Indonesia. From these observations they deduced a simple arithmetical rule: the number of species approximately doubles with every tenfold increase in area. The qualitative relationship between area and number of species—the bigger the area, the more the species—seems intuitively obvious; and the quantitative relationship derives from empirical observation.

Though simple—even simplistic—the theory seems robust. Nevertheless, a rigorous test of the theory would make it more valuable, and this is precisely what Lovejoy set out to perform with his Brazilian rainforest experiment. Destined to continue for many more decades, the experiment has already produced sufficient information to put to rest any serious doubts about the theory's central premise.

There are many ways in which the actual number of species in a habitat of a certain size may be influenced up or down, of course. A thousand acres of flat terrain are likely to support fewer species than a thousand acres of extremely varied topography, for instance. The reason is that many more microhabitats are present in the latter than the former. And a thousand tropical acres will support more species than a similar area at high latitudes, for reasons I discussed in chapter 7. As long as appropriate comparisons are made—that is, similar latitudes, similar terrain—island biogeography theory is a powerful tool for making predictions. It is also the only tool, aside from counting species one by one; that is usually not practical. When Julian Simon says that Wilson's mathematical model "is based on nothing but speculation"[12] and dismisses predictions as "the statistical flummery of species loss,"[13] he is being willfully ignorant of the facts underlying the theory.

Armed with this tool, what can we say about the consequences of reducing tropical forests to 10 percent of their original extent? The arithmetical relationship based on the theory predicts that 50 percent of species will go extinct—some immediately, some over a period of decades or even centuries. If most ecologists accept this empirical relationship as a reasonable guide, why are estimates of projected species extinction over the next century so much at variance with one another? Why does one authority state that 17,000 species will be lost every year while another puts the figure at 100,000?

The reasons are several, not the least of which is a great uncertainty about how many species exist in the world. As I said in chapter 7, estimates range from ten million to a hundred million. Using the same 50 percent proportion for species loss, therefore, one person using the higher estimate will produce an absolute number that is an order of magnitude greater than one who elects to use the lower estimate. There are other confounding factors, too, such as great (and unknown) differences in the size of habitat fragments that escape destruction, and uncertainties in the ranges of most species. If, for example, a significant proportion of species is restricted to small localities, then the loss of species will be higher than 50 percent, and may approach the percentage of habitat lost. "That there is considerable spread in the estimates is really not surprising, given the difficulties in getting precise information," comments Lovejoy. He then adds the key to this argument: "What is important is that every effort to estimate rates has produced a *large* number."[14] Few dispute the proportion of species destined to disappear if current trends continue—that is, something close to half. Fifty percent of the total of the world's species is a large number.

Even if we take a figure in the lower range of estimates, say thirty-thousand species per year, the implication is still startling. David Raup has calculated from the fossil record that during periods of normal, or background, extinction, species loss occurs at an average of one every four years. Extinction at the rate of thirty-thousand a year, therefore, is elevated 120,000 times above background. This is easily comparable with the Big Five biological crises of geological history, except that this one is not being caused by global temperature change, regression of sea level, or asteroid impact. It is being caused by one of Earth's inhabitants. *Homo sapiens* is poised to become the greatest catastrophic agent since a giant asteroid collided with the Earth sixty-five million years ago, wiping out half the world's species in a geological instant.

•   •   •   •

The figures I've been talking about are predictions for extinction rates early in the next century if current trends of habitat destruction continue. Critics not only doubt the validity of these predictions, but also challenge ecologists to produce hard evidence of an alarming level of human-caused extinctions today. It is true that, because there has been no comprehensive, global survey, ecologists are unable to proffer such evidence in the form of a complete list of extinctions. In effect, however, the critics are implying that no such evidence exists because no (or very few) species are disappearing as a result of human activity. Despite the lack of a comprehensive survey, there is a large body of isolated studies in many different habitats around the world. Dismissed by the critics as "merely anecdotal," these studies collectively give more than enough reason for concern.

I will offer some examples. I've already mentioned the massive loss of fish species in Lake Victoria. By itself, the disappearance of two-hundred species in twenty years is already way beyond the background extinction rate of one species every four years. If background extinction rates applied to birds, for instance, ecologists should expect to see the disappearance of a bird species no more frequently than once every century. And yet, as Stuart Pimm reports, "In the Pacific alone, we are seeing about one extinction per year."[15] Pimm's field work is in Hawaii, where birds are his special interest. The Hawaiian islands may look like a tropical paradise to tourists, but to ecologists they bear the scars of recent, catastrophic extinctions. As many as half the islands' bird species have gone extinct since first human contact, and the loss continues today. Of some 135 bird species there, only eleven thrive in numbers that ensure their survival well into the next century. "A dozen . . . are so rare that there is little hope of saving them," says Pimm. "A further dozen are legally classified as Endangered—meaning that their future survival is uncertain."[16]

A little more than a decade ago, ninety species of plants became extinct in a virtual instant, when the forested ridge on which they grew was cleared for agricultural land. The ridge, in the western Andean foothills of Ecuador, is called Centinela, and among ecologists the name has become synonymous with catastrophic extinction at human hand. By chance, two ecologists, Alwyn Gentry and Calaway Dodson, visited the ridge in 1978 and carried out the first botanical survey in its cloud forest. Among the riot of biodiversity that is nurtured by this habitat, Gentry and Dodson discovered, were ninety previously unknown species, including herbaceous plants, orchids, and epiphytes, that lived nowhere else. Centinela was an ecological island, which, being isolated, had developed a unique flora. Within eight years the ridge had been transformed into farmland, and its endemic species were no more.

Centinela had a unique flora, but it wasn't unique in being an ecological island. Countless such ridges exist along the whole length of the Andes, most of which, too, must have developed species not found elsewhere. What made the Centinela habitat notorious was that a botanical survey had been carried out prior to its destruction. Each time an ecological island is cleared, species will vanish in a virtual instant, an event ecologists now term a *centinelan extinction*. There are two points to be emphasized here. The first is that whenever ecologists are able to survey a habitat before and after disturbance, species loss is almost always seen, often a catastrophic one. However, in the vast majority of instances, habitat destruction occurs in areas that have not been surveyed for their flora and fauna, so it is more than likely that countless species become extinct before ecologists even know of their existence. How is one to document this, except by extrapolation? The second is that, like the plants on Centinela, many species have very limited ranges, particularly in the tropics, so destruction of habitat often results in the instant destruction of species. As I indicated earlier, this implies that the 50 percent figure predicted for eventual species loss is more likely to be an underestimate than an overestimate.

The list of "anecdotal" evidence is long: half the freshwater fish of peninsular Malaysia, ten bird species of Cebu in the Philippines, half the forty-one tree snails in Oahu, forty-four of the sixty-eight shallow-water mussels of the Tennessee River shoals, and so on. The evidence may be anecdotal in the sense of its not being the result of a systematic survey, but it is compelling nonetheless. In an attempt to be quantitative with the known extinction data, and thereby come up with an assessment of whether or not we face a biological crisis of our own making, Stuart Pimm and two of his colleagues analyzed some of the best known and most closely documented cases. These include freshwater mussels and freshwater fish in North America, mammals in Australia, plants in South Africa, and amphibians worldwide. "What causes extinction?" Pimm and the others ask rhetorically. "Our reading of the five case studies is that species introductions and physical habitat alteration are the highest-ranking factors."[17] I won't go into the details of the recorded extinctions, because they can be found in Pimm's publication; instead, I'll concentrate on the conclusions that flow from the analysis of them.

If the observed levels of extinction known in these cases is typical for similar species worldwide, then current extinction is running at a rate some thousand to ten-thousand higher than background extinction. Skeptics may argue that these examples represent particularly high levels of extinction, and are therefore not representative. Even if this is the case, say Pimm and his colleagues, and these known extinctions are the only ones in these groups of species worldwide, which is highly improbable, then the rate is *still* two-hundred to a thousand higher than background. This qualifies as a mass extinction. The authors point out that none of the cases is from areas where human densities are particularly high, illustrating that the hand of death is effective at a distance. How much more effective would it be, then, in the midst of high concentrations of humanity? Pimm asks what we are to conclude from this and other studies: "Those who suggest that high extinction rates are a fabrication seem curiously ignorant of the facts."[18] Or, perhaps, willfully ignorant.

The documentation of known extinctions may seem to be the only way to demonstrate that we are in the midst of a biotic crisis, and this is what skeptics demand. After all, there can be no case for murder without a body. Equally, if a population of a species exists somewhere, it is not extinct, is it, even if its total range is reduced by habitat destruction? However, this point of view underestimates both the magnitude of the current crisis and its complexity. "It is important to recognize that, except when all individuals of a species are simultaneously eliminated, as by a meteor or hurricane, extinction is a multi-stage process,"[19] observes Daniel Simberloff. By way of example, he cites the case of the heath hen, which I recounted in chapter 5. The cause of extinction is usually given as hunting and habitat destruction by humans. The bird's range, remember, was huge, and covered much of the eastern seaboard of the United States. Hunting and habitat destruction reduced the species' number to fifty individuals in 1908, when a reserve was established to save it from extinction. Over the next two decades the population's numbers began to rise robustly, but eventually the species did go extinct, through a combination of biblical calamities, including fire and pestilence.

The point of the story is that once the heath hen population was reduced to small numbers, its eventual extinction was virtually assured. As I've stated several times, a small population is vulnerable to normal fluctuations in its numbers, the consequence of disease and disasters. A population of a thousand individuals can weather a population drop of a hundred; such a fluctuation spells the end for a population that starts with only a hundred individuals. In the case of the heath hen, even when hunting and habitat alteration were halted, its survival was precarious in the extreme. A proper assessment of the impact of human activity on current biodiversity therefore must take into account populations that have become so small as to be likely victims to stochastic fluctuations or are trending in that direction. This is precisely what Stuart Pimm did in describing the prospects of the Hawaiian birds. Only eleven are assured of survival well into the next century. Populations of the remaining 124 species have already been reduced, in some cases perilously so. And yet a simple species accounting notes that 135 species exist: no extinction to report. Simberloff describes the predicament graphically: "Many populations, including the last populations of some species, might be superficially healthy but among the living dead."[20]

I believe that the "anecdotal" accounts of extinctions worldwide that ecologists are currently telling us about are but the merest hint of a catastrophic reality that is unfolding silently and, for the most part, away from our sight. Given the absolute impossibility of documenting the demise of every species whose fate is sealed by human activity, we need to be acutely sensitive to these faint echoes on the wind, because they carry an important message. Dominant as no other species has been in the history of life on Earth, *Homo sapiens* is in the throes of causing a major biological crisis, a mass extinction, the sixth such event to have occurred in the past half billion years. And we, *Homo sapiens,* may also be among the living dead.

14

Does It Matter?

PAUL EHRLICH has an analogy for those who contend that, because ecologists cannot say precisely how many species are endangered, it is premature to be alarmed about the putative impending collapse of biodiversity. "[It is like] saying that people should not be overly concerned about the burning down of the world's only genetic library because the number of 'books' in it is not known to within an order of magnitude, and fire modelers disagree on whether it will be half consumed in a couple of decades or whether that level of destruction might take fifty years," he wrote recently in a letter to *Science*. "Apparently a few scientists would never call the fire department unless they could inform it of the exact temperature of the flames at each point in a holocaust."[1]

I have my own analogy. Imagine that a giant asteroid is spotted on collision course with Earth. Many people would be justly worried, because such impacts are thought to have unleashed mass extinctions in the past. My reading of the logic of Julian Simon and his ilk is that they would argue there is no cause for alarm, because theories of mass extinction as a result of asteroid impact are pure speculation and guesswork; no one has ever seen such an extinction happen; and, anyway, the asteroid might miss. If there were some means of deflecting the asteroid, the cost of not doing so would be catastrophic, in the event that the Simon view turned out to be wrong. What are the costs of his being wrong about the sixth extinction? What would it matter to us and to the rest of the world's biota if half its species were to be pushed into oblivion sometime during the next century?

There are several answers to these questions, depending on the time frame in which they are cast. One of them is that, in the long run, it matters not at all. Although true in a sense, and many anti-alarmists adduce it in their support, I will argue that such a response reflects an ignorance of patterns in the history of life and our place in them.

In chapter 8 I talked about the value of biodiversity, and identified three important areas: economic, ecosystem services, and esthetic. I won't go into this in any detail again here, except to say that where value can be identified, it follows that the loss of diversity represents the loss of that value. If animals and plants are a potential source of new materials, new foods, and new medicines, then the loss of species reduces that potential. If an interacting network of plants and animals is important in sustaining the chemistry of the atmosphere and the soil, the loss of species reduces the efficacy of these services. And if a rich diversity of species succors the human psyche in important ways, then the loss of species reduces us in some ineffable way. A legitimate question in each of these three areas, however, is the following: Are all the existing species necessary for satisfying economic worth, ecosystem services, and esthetic pleasure? Or can some be lost without harm?

To Julian Simon, the answer is obvious, and he displays it in reference to the loss of species that occurred when settlers clear-cut the Middle West of the United States. "It seems hard to even *imagine* that we would be enormously better off with the persistence of any such imagined species," he suggested in his debate with Norman Myers. "This casts some doubt on the economic value of species that might be lost elsewhere."[2] Simon's principal measures of value are economics and immediate practicality, as illustrated by a further remark he made in the debate. "Recent

scientific and technical advances—especially seed banks and genetic engineering—have diminished the importance of maintaining species in their natural habitat."[3] I'll contrast this with a
very different way of looking at the value of species in their natural habitat, made by Les Kaufman in a chapter in his book *The
Last Extinction:* "A piece of the American soul died along with the
passenger pigeon, plains buffalo, and American chestnut."[4] Although I would not argue that each and every species in the
world *must* be saved, especially at the cost of maintaining human
welfare, my sentiment is closer to Kaufman's than to Simon's.
Humans evolved within a world of nature, and an appreciation
of, and *need* for, nature are real and ineradicable components of
the human psyche. We risk eroding the human soul if we allow
the erosion of the richness of the world of nature around us.

But suppose the human psyche could be nurtured through the
trauma of an ecologically impoverished world. Suppose, too, that
present and future technologies could provide us with all the
material resources that we presently and potentially derive from
the natural world. Would we be able "to feed, clothe, and supply
energy to an ever growing population for the next seven billion
years," as Julian Simon contends? It is true that throughout human history the material quality of life has steadily increased,
even as the size of the population has increased. Guided by this
history, Simon assumes that the same pattern can extend essentially indefinitely into the future, and that there is no limit to
what humans can take from the world of nature without detriment to ourselves or to nature itself. In other words, he believes
that our continued appropriation of nature is compatible with
sustaining an equable natural world. History is, of course, a use-
ful guide to the future, but it can also blind us to emerging
realities. Science and technology have increased our creature
comforts, of that there is no doubt, but those comforts may blind
us to the reality of the global environment. Nurtured as many of
us are in our artificial urban environments, we do not see the
relationship between the inputs and outputs of the natural econ-
omy of the Earth.

The inputs and outputs are the interactions among species at
all scales of life, from filaments of fungi nurturing the health of
plant rootlets to the global chemical cycles of water, oxygen, and
carbon dioxide. They are the ecosystem services to which I re-

ferred earlier, and they represent the tangible elements of the stability and health that emerge from the entire biota of the Earth operating as a complex dynamic system. How exactly do health and stability emerge? We don't know. Can the system be reduced in size, through eliminating a proportion of species in all ecological realms, and still be effective? We don't know. Which are its most important components? We know this only incompletely. Which species or groups of species can be removed without detriment to the system that sustains us and all living organisms? Again, we know this only incompletely. The degree of ignorance about the natural world upon which we depend is frustratingly large, but it is not total, as I described in chapter 8. We do know that *Homo sapiens* is not exempt from the rules that govern the lives of all other organisms.

In the face of ignorance about how much of current biodiversity we need in order to sustain a healthy Earth's biota, is it more responsible to say (1) because we don't know if we need it all, we can safely assume we don't; or (2) we recognize the complexities of the system, and assume we do? The answer is obvious, because the costs of being wrong on the first count are enormous. In any case, many ecologists, extrapolating from the incomplete knowledge they have about the structure and dynamics of ecosystem services, believe that we do need all, or at least most, of what we currently have. Through continued destruction of biodiversity in the wake of economic development, we could push the natural world over a threshold beyond which it might be unable to sustain, first, itself and, ultimately, us. Unrestrained, *Homo sapiens* might not only be the agent of the sixth extinction, but also risks being one of its victims.

Humans live in the present. We look at the world around us and find it difficult to encompass change over great tracts of time. But the perspective of time is important if we are fully to understand the biological processes we are driving by our actions, and, of course, to see where our future as a species lies. We must therefore turn to the fossil record of life, for it alone can inform us of the dynamics of living systems at time scales beyond our current experience and imagination.

The most immediate message of the record about the history of life is that major catastrophic collapses of biological diversity can and do occur. Moreover, these crises in life's flow can be rapid, irreversible, and unpredictable. This should press home to us an important insight into the natural world of which we are a part: species and communities of species are not infinitely resilient to external insult; they are vulnerable, and they can disappear, to be lost forever. We acknowledge that mass extinctions can be precipitated by Earth's impact with extraterrestrial objects and by global changes of various kinds, but do not see ourselves as a potential agent in such biological crises. The daily cutting of tropical forests and encroachment on wild habitats is a less dramatic process than asteroid impact, but in the end the effect is the same. Insidiously, a mass extinction is occurring. In pursuing our own ends, we treat the world of nature as if it can withstand each of our assaults without harm, but we do so at our peril.

The fossil record shows us that life has not been a static phenomenon through Earth history, but rather is a dynamic process. Neither is it a steady progression, but rather is punctuated by mass killings, the victims of which—whether we are looking at individual species or groups of species or ecosystems—are gone forever. The death of a species is the termination of a continuous chain of genetic links that reaches back billions of years; a unique genetic package vanishes from Earth's variety for good. Each time human action results in the extirpation of a species, collectively each of us bears a part of the responsibility for snuffing out a unique part of life, forever. I take this responsibility very seriously.

But, retort the anti-alarmists, look at the fossil record again and you will see that species' lives are limited anyway, lasting between one and ten million years on average. (The longer-lived species are among the less conspicuous creatures of our world; those with shorter life spans are the larger creatures, such as terrestrial vertebrates.) Some species end their tenure in the steady grind of background extinction, others in the cataclysmic mass dyings. With this perspective, say the anti-alarmists, attempts to save species "may be wasting time, effort, and money [because they] will disappear over time, regardless of our efforts."[5] As Stephen Jay Gould correctly retorts, this viewpoint "makes about as much sense as arguing that we shouldn't treat an easily curable childhood infection because all humans are ultimately mortal."[6] I'll return to this shortly, because it contains a vital element of an ethical argument that encompasses geological time and our place in it.

The second major message of the fossil record is that evolution is a wondrously and powerfully creative process, one that rapidly fills the void left after each mass extinction. After all, the diversity of life in recent times is at an all time high, the result of repeated bouncebacks after five major biological crises and more than a dozen smaller ones. Often, the burgeoning of new species after extinction events involves a transformation of the dominant form of life. We are in the age of mammals, which came into being after the demise of the dinosaurs, sixty-five million years ago. And in this new age, primates have become the most extensively endowed mentally, with *Homo sapiens* the top of that heap. After the sixth extinction is over, diversity will return, as it always has, assuming of course that the agent of destruction—the current behavior of *Homo sapiens*—passes. And, if the past is anything to judge by, the diversity of life will be even more extensive than it is now. And who knows what evolutionary novelties may emerge? If nature recovers so boisterously following mass extinctions, then perhaps we shouldn't be concerned about causing one. The answer to this is that it depends on the time scale you are looking at.

Mass extinctions are virtually instantaneous, occurring within a matter of years or centuries in the case of asteroid impact to millennia or a few million years from Earth-bound causes. Recovery, however, is slow, lasting somewhere between five and twenty-five million years. Slow, that is, on a human scale. Slow not only in terms of time as we can comprehend it as individuals, but also in terms of the expected tenure of us as a species.

There is no reason to think that the one- to ten-million-year average life span that applies to other species should not apply to our own. *Homo sapiens* has been around for perhaps 150,000 years, so we might look forward to a further million years or so (being a large terrestrial vertebrate), unless, of course, our capacity for destruction hastens our end. Moreover, at some point in Earth's future, a giant asteroid or comet will slam into our planet, instantly extirpating the majority of species, perhaps including our own. Several small asteroids have passed uncomfortably close in recent years. They are harbingers of the inevitable, which, according to some calculations, will arrive about thirteen million years from now. If, by an unusual chance, the descendants of *Homo sapiens* still thrive on Earth at that time, the aftermath of the impact would surely extirpate most, if not all, of their populations; and even if some do survive the initial impact, civilization would certainly be shattered, perhaps never to recover. If there is one certainty that we can derive from an understanding of life's flow and the forces that shape it, it is that one day we and our descendants will be no more, and the Earth and its inhabitants will go on without us.

Many people find it impossible to contemplate a time when *Homo sapiens* would no longer exist, so they like to assume that we will break the biological rule and continue forever, or at least until our planet ceases to exist, billions of years from now, when its atmosphere is burned off by an expanding sun. Julian Simon obviously believes so, when he talks of our capacity to thrive for the next seven billion years. He is blind to reality. There are those who cling to the idea that we can escape this end, too, by recourse to space travel and colonization of other planets, in which case it matters little what damage we inflict on the one planet we know can support us. Both are flights of fancy, born of the arrogant belief that *Homo sapiens* is separate from and above the rest of the world of nature, and a belief in our invincibility. If we learn anything from a scrutiny of life's history and of the dynamics by which species thrive collectively, we learn that neither is true.

I could argue, as others have, that we owe it to ourselves, to our children, and to our children's children not to foul our nest, not to degrade the wondrous diversity of life upon which we depend for our survival and our soul. I could argue, as others have, that, as the one sentient creature on Earth, we have a duty to protect the lives of all of Earth's species. I could argue, as others have, that all the species with which we share the world today have an absolute right to our protection, simply because they exist. Each of these exhortations is valid, and separately or collectively they represent an imperative to recognize our role in the sixth extinction and to halt the insidious mass destruction that is soon to push one hundred species a day, four species an hour, into evolutionary oblivion. But I want to add a further imperative, one that derives from the perspective of the history of life on Earth, not, as earlier, from a human-centered view, but from the view of the rest of nature's species.

The sixth extinction is similar to previous biological catastrophes in many ways. For instance, the most vulnerable species are those whose geographical distribution is limited, those in and near the tropics, and those with a large body size. It is unusual in several ways, too, most particularly in that large numbers of plant species are being wiped out, which is unprecedented compared with past crises. But in the end, with the passage of five, ten, or twenty million years, despite this and other distortions of the biota that will remain, rebound will occur. "On geological scales, our planet will take care of itself and let time clear the impact of any human malfeasance," as Gould has put it.[7] Why, then, if it matters not at all in the long run what we do while we are here, should we concern ourselves with the survival of species that, like us, will eventually be no more?

We should be concerned because, special though we are in many ways, we are merely an accident of history. We did not arrive on Earth as if from outer space, set down amid a wondrous diversity of life, blessed with a right to do with it what we please. We, like every species with which we share the world, are a product of many chance events, leading back to that amazing explosion of life forms half a billion years ago, and beyond that to the origin of life itself. When we understand this intimate connection with the rest of nature in terms of our origins, an ethical imperative follows: it is our duty to protect, not harm, them. It is our duty, not because we are the one sentient creature on Earth, which bestows some kind of benevolent superiority on us, but because in a fundamental sense *Homo sapiens* is on an equal footing with each and every other species here on Earth. And when we understand the Earth's biota in holistic terms—that is, operating as an interactive whole that produces a healthy and stable living world—we come to see ourselves as part of that whole, not as a privileged species that can exploit it with impunity. The recognition that we are rooted in life itself and its well-being demands that we respect other species, not trample them in a blind pursuit of our own ends. And, by this same ethical principle, the fact that one day *Homo sapiens* will have disappeared from the face of the Earth does not give us license to do whatever we choose while we are here.

If what I've been saying sounds idealistic, I will admit to it. My dual careers as paleontologist and conservationist have given me a unique view on the value of life's diversity and the way it changes through time. But I also have a practical view, which derives from seeing people struggle to survive by exploiting the only resources available to them—namely, the natural world around them. Aiding their struggle while halting the destruction of those resources is humanity's greatest challenge for the next century. It can be done, but only if the different needs of rich and poor countries are acknowledged. It will fail if the richer nations try to impose solutions that effectively freeze the citizens of the less-developed nations in permanent poverty.

For a long time mass extinctions were a neglected subject of study, because they were mysterious in many ways and, anyway, were thought to be mere interruptions in the flow of life. They are now recognized as a major creative force in shaping that flow, and they will surely continue to be so for billions of years into the future, long after *Homo sapiens* and its descendants are no more. But much of the mystery of mass extinctions remains; specifically, what exactly causes them. As David Raup wrote in his book *Extinction: Bad Genes or Bad Luck?:* "The disturbing reality is that for none of the thousands of the well-documented extinctions in the geological past do we have a solid explanation of why the extinction occurred."[8] For each of the Big Five there are theories of what caused them, some of them compelling, but none proven.

For the sixth extinction, however, we do know the culprit. We are.

Notes

CHAPTER 2 Life's Salient Mystery

1. Charles Darwin, *Origin of Species* (London: Penguin Books, 1985 edition), p. 312.

2. Andrew H. Knoll, "End of the Proterozoic Eon," *Scientific American*, October 1991, p. 64.

3. Darwin, *op. cit.*

4. Adolf Seilacher, "Precambrian evolutionary experiments: Vendozoa and Psammocorallia," in Pere Alberch and Gabriel A. Dover, eds., *The Reference Points in Evolution*, Fundancion Juan March, No. 255, Serie Universitaria, 1990, p. 480.

5. *Ibid.*, p. 49.

6. Simon Conway Morris, "Burgess Shale faunas and the Cambrian explosion," *Science*, vol. 246 (1989), p. 339.

7. *Origin and Evolution of Metazoa*, Jerr H. Lipps and Philip W. Signor, eds. (New York: Plenum Press, 1992), p. 17.

CHAPTER 3 The Mainspring of Evolution

1. David Jablonski and David J. Bottjer, "The ecology of evolutionary innovation: the fossil record," in *Evolutionary Innovations*, M. H. Nitecki, ed. (Chicago: University of Chicago Press, 1990), p. 253.

2. Cited in Carol Kaesuk Yoon, "Biologists' 'Big Bang' took a mere blink of an eye," *New York Times*, Science section, page 1, 7 September 1993.

3. Jablonski and Bottjer, *op. cit.*

4. Jeffrey S. Levinton, "The Big Bang of animal evolution," *Scientific American*, November 1992, p. 84.

5. Stephen Jay Gould, "Treasures in a taxonomic wastebasket," *Natural History*, December 1985, p. 32.

6. Douglas Erwin *et al.*, "A comparative study of diversification events: the early Paleozoic versus the Mesozoic," *Evolution*, vol. 41 (1987), p. 1177.

7. Edward O. Wilson, *The Diversity of Life* (New York: W. W. Norton, 1992), p. 192.

8. C. D. Walcott, cited in Simon Conway Morris and Harry B. Whittington, "The animals of the Burgess Shale," *Scientific American*, July 1979, p. 122.

9. Harry Whittington, cited in Roger Lewin, *Thread of Life* (Washington D.C.: Smithsonian Books, 1982), p. 107.

10. Simon Conway Morris and Harry B. Whittington, *op. cit.*, p. 131.

11. Stephen Jay Gould, "Play it again, life," *Natural History,* February 1986, p. 20.

12. Charles Darwin and Alfred Russel Wallace, "On the tendency of species to form varieties," paper read at the Linnean Society, 1 July 1858.

13. Simon Conway Morris and Harry B. Whittington, *op. cit.*, p. 132.

14. Stephen Jay Gould, *Wonderful Life* (New York: W. W. Norton, 1989), p. 237.

15. ———, "In touch with Walcott," *Natural History,* July 1990, p. 12.

16. Simon Conway Morris, "Burgess Shale faunas and the Cambrian explosion," p. 339.

17. Stephen Jay Gould, "Play it again, life," *Natural History,* p. 24.

18. Simon Conway Morris, "Burgess Shale faunas and the Cambrian explosion," p. 345.

19. Jeffrey S. Levinton, *op. cit.*, p. 91.

CHAPTER 4   **The Big Five**

1. Charles Darwin, *op. cit.*, p. 322.

2. *Ibid.*, p. 321.

3. *Ibid.*

4. *Ibid.*, p. 323.

5. *Ibid.*

6. David M. Raup, "Diversity crises in the geological past," in *Biodiversity,* E. O. Wilson, ed. (Washington D.C.: National Academy Press, 1988), p. 52.

7. ———, *Extinction: Bad Genes or Bad Luck?* (New York: W. W. Norton, 1991), pp. 112–113.

8. Paul Wignall, "The day the world nearly died, *New Scientist,* 25 January 1992, p. 55.

9. Douglas H. Erwin, "The Permo-Triassic Extinction," *Nature,* vol. 367 (1994), p. 231.

10. Steven M. Stanley, *Extinction* (New York: Scientific American Library Books, 1987), p. 40.

11. Stephen Jay Gould, "The Cosmic Dance of Siva," *Natural History,* August 1984, p. 16.

12. David M. Raup, "Changing views of natural catastrophe," *The Great Ideas of Today* (Encyclopedia Britannica, Inc., 1988), p. 55.

13. William Clemens, cited in Richard A. Kerr, "Huge Impact Tied to Mass Extinction," *Science,* vol. 257 (1992), p. 879.

14. Anthony Hallam, cited in Richard A. Kerr, *op. cit.*, p. 880.

CHAPTER 5   **Extinction: Bad Genes or Bad Luck?**

1. Charles Darwin, *op. cit.*, p. 343.

2. *Ibid.*, p. 119.

3. Stephen Jay Gould, "The Cosmic Dance of Siva," p. 17.

4. *Ibid.*, p. 18.

5. David Jablonski, "Causes and consequences of mass extinction," in *Dynamics of Extinction*, D. K. Elliot, ed. (New York: John Wiley and Sons, 1986), p. 209.

6. David M. Raup, "Extinction: bad genes or bad luck?" *New Scientist*, 14 September 1991, p. 47.

7. ———, "The role of extinction in evolution," ms. for the *Proceedings of the National Academy of Sciences*, in press, p. 18.

8. Stephen Jay Gould, "The Cosmic Dance of Siva," p. 18.

9. David Jablonski, "Background and mass extinctions: the alternation of macroevolutionary regimes," *Science*, vol. 231 (1986), p. 131.

10. J. John Sepkoski, Jr., "Phylogenetic and ecologic patterns in the Phanerozoic history of marine biodiversity," in *Systematics, Ecology, and the Biodiversity Crisis*, Niles Eldredge, ed. (New York: Columbia University Press, 1992), p. 84.

CHAPTER **6** *Homo sapiens*, the Pinnacle of Evolution?

1. Charles Darwin, *op. cit.*, p. 459.

2. Alfred Russel Wallace, "The Limits of Natural Selection," in *Essays on Natural Selection* (London: Macmillan, 1871), p. 359.

3. Robert Broom, *The Coming of Man: Was It Accident or Design?* (London: Witherby, 1933), p. 220.

4. Sir Arthur Keith, *A New Theory of Human Evolution* (London: The Philosophical Library, 1949), p. 161.

5. Sir Grafton Elliot Smith, *Essays on the Evolution of Man* (Oxford: Oxford University Press, 1923), p. 1.

6. Julian Huxley, "Evolutionary processes and taxonomy with special reference to grades," in *Systematics of Today*, Olav Hedberg, ed. (Uppsala: Uppsala Universitets Arsskrift, 1986), p. 35.

7. Stephen Jay Gould, "Play it again, life," *Natural History*, p. 19.

8. ———, *Wonderful Life*, p. 318.

9. *Ibid.*, p. 291.

10. ———, "Eternal metaphors of paleontology," in *Patterns of Evolution as Illustrated in the Fossil Record*, Anthony Hallam, ed. (New York: Elsevier, 1977), p. 13.

11. ———, *Ever Since Darwin* (New York: W. W. Norton, 1977), p. 45.

12. Charles Darwin, *op. cit.*, p. 337.

13. George Gaylord Simpson, *The Meaning of Evolution* (New Haven: Yale University Press, 1949), p. 252.

14. ———, *Principles of Animal Taxonomy* (New York: Columbia University Press, 1961), p. 97.

15. Stephen Jay Gould, cited in Roger Lewin, *Complexity: Life at the Edge of Chaos* (New York: Macmillan, 1992), pp. 145–146.

16. Edward O. Wilson, *The Diversity of Life*, p. 187.

CHAPTER 7   Endless Forms Most Beautiful

1. Charles Darwin, *op. cit.*, p. 460.

2. Michael A. Mares, "Neotropical mammals and the myth of Amazonian biodiversity," *Science*, vol. 255 (1992), p. 976.

3. Stuart L. Pimm and John L. Gittleman, *Science*, vol. 255 (1992), p. 940.

4. George Stevens, cited in "Biologists disagree over bold signature of nature," *Science*, vol. 244 (1989), p. 527.

5. Michael A. Rex *et al.*, "Global-scale latitudinal patterns of species diversity in the deep-sea benthos," *Nature*, vol. 365 (1993), p. 636.

6. David Jablonski, "The tropics as a source of evolutionary novelty through geological time," *Nature*, vol. 364 (1993), p. 142.

7. Wallace Arthur, "The bulging biosphere," *New Scientist*, 29 June 1991, p. 43.

8. John D. Gage and Robert M. May, "A dip into the deep seas," *Nature*, vol. 365 (1993), p. 610.

9. Robert M. May, "Biological diversity: differences between land and sea," *Transactions of the Royal Society*, Series B, vol. 343 (1994), p. 109.

10. *Ibid.*

11. Robert M. May, "How many species are there?" *Science*, vol. 241 (1988), p. 1441.

12. Edward O. Wilson, "The biological diversity crisis: a challenge to science," *Issues in Science and Technology*, Fall 1985, p. 22.

13. Robert M. May, "How many species inhabit the Earth?" *Scientific American*, October 1992, p. 48.

14. Edward O. Wilson, "The biological diversity crisis: a challenge to science," *Issues in Science and Technology*, Fall 1985, p. 230

CHAPTER 8   Value in Diversity

1. David Ehrenfeld, "Why put value on biodiversity?" in *Biodiversity*, E. O. Wilson, ed., p. 212.

2. *Ibid.*, p. 213.

3. Charles M. Peters *et al.*, "Valuation of an Amazonian rainforest, *Nature*, vol. 339 (1989), p. 656.

4. Hugh H. Iltis, "Serendipity in the exploration of biodiversity," in *Biodiversity*, E. O. Wilson, ed., pp. 102–103.

5. David Ehrenfeld, "Why put value on biodiversity?" in *Biodiversity*, E. O. Wilson, ed., p. 214.

6. *Ibid.*, p. 213.

7. Charles Darwin, *op. cit.*, p. 459.

8. James E. Lovelock, "The Earth as a living organism," in *Biodiversity*, E. O. Wilson, ed., p. 488.

9. David Tilman and John A. Downing, "Biodiversity and stability in grasslands," *Nature*, vol. 367 (1994), p. 365.

10. Edward O. Wilson, "Biophilia and the conservation ethic," in *The Bio-*

philia Hypothesis, Stephen R. Kellert and Edward O. Wilson, eds. (Island Press: Washington D.C., 1993), p. 31.

11. Luther Standing Bear, *Land of the Spotted Eagle* (Lincoln: University of Nevada Press, 1933), p. 45.

CHAPTER 9   **Stability and Chaos in Ecology**

1. Seth R. Rice, "Nonequilibrium determinants of biological community structure," *American Scientist,* vol. 82 (1994), p. 427.

2. Stuart L. Pimm, *The Balance of Nature* (Chicago: University of Chicago Press, 1991), p. 4.

3. Robert May, "The chaotic rhythms of life," *New Scientist,* 18 November 1989, p. 37.

4. *Ibid.,* p. 39.

5. Alan Hastings and Kevin Higgins, "Persistence of transients in spatially structured ecological models," *Science,* vol. 263 (1994), p. 1136.

6. William W. Schaffer and Mark Kot, "Chaos in ecological systems," *Trends in Ecology and Evolution,* vol. 1, September 1986, p. 63.

7. Michael P. Hassell *et al.,* "Species coexistence and self-organizing spatial dynamics, *Nature,* vol. 370 (1994), p. 290.

8. Ted J. Case, "Invasion resistance arises in strongly interacting species-rich model competition communities," *Proceedings of the National Academy of Sciences,* vol. 87 (1990), p. 9610.

9. *Ibid.*

10. Stuart L. Pimm *et al.,* "Food web patterns and their consequences," *Nature,* vol. 350 (1991), p. 669.

11. Jeremy B.C. Jackson, "Community unity?" *Science,* vol. 264 (1994), p. 1412.

12. Brian Walker, "Diversity and stability in ecosystem conservation," in David Western and Mary Pearl, eds., *Conservation for the Twenty-first Century* (New York: Oxford University Press, 1989), p. 125.

13. *Ibid.,* p. 130.

CHAPTER 10   **Human Impacts of the Past**

1. Alfred Russel Wallace, *The Geographical Distribution of Animals,* vol. 1 (New York: Harper and Brothers, 1876), p. 151.

2. Alfred Russel Wallace, *The World of Life* (New York: Moffat, Yard, 1911), p. 264.

3. Richard Owen, *Paleontology, or a Systematic Study of Extinct Animals and Their Geological Relations* (Edinburgh: A. and C. Black, 1860), p. 399.

4. Charles Lyell, *A Second Visit to the United States of North America,* vol. 1 (New York: Harper and Brothers, 1849), p. 259.

5. Paul S. Martin, "Prehistoric overkill: the global model," in *Quaternary Extinctions: a prehistoric revolution,* Paul S. Martin and Richard G. Klein, eds. (Tucson: University of Arizona Press, 1984), p. 367.

6. *Ibid.,* p. 370.

7. John E. Guilday, "Pleistocene Extinction and Environment Change," in *Quaternary Extinctions: a prehistoric revolution,* Paul S. Martin and Richard G. Klein, eds., p. 250.

8. *Ibid.,* p. 254.

9. *Ibid.,* p. 257.

10. Paul S. Martin, "Prehistoric overkill: the global model," in *Quaternary Extinctions: a prehistoric revolution,* Paul S. Martin and Richard G. Klein, eds., p. 384.

11. *Ibid.,* p. 323.

12. Jared Diamond, *The Third Chimpanzee* (New York: HarperCollins, 1992), p. 321.

13. *Ibid.*

14. *Ibid.*

15. A. Berger, *Hawaiian Birdlife* (Honolulu: University Press of Hawaii, 1972), p. 7.

16. Helen F. James and Storrs L. Olson, "Flightless birds," *Natural History,* September 1983, p. 30.

17. *Ibid.*

18. Storrs L. Olson and Helen F. James, "The role of Polynesians in the extinctions of the avifauna of the Hawaiian Islands," in *Quaternary Extinctions: a prehistoric revolution,* Paul S. Martin and Richard G. Klein, eds., pp. 777–778.

19. David Tilman *et al.,* "Habitat destruction and the extinction debt," *Nature,* vol. 371 (1994), p. 66.

20. Storrs L. Olson, "Extinction on islands: man as a catastrophe," in *Conservation for the Twenty-First Century,* David Western and Mary Pearl, eds. (New York: Oxford University Press, 1989), p. 52.

CHAPTER 11 **The Modern Elephant Story**

1. Douglas H. Chadwick, *The Fate of the Elephant* (San Francisco: Sierra Club Books, 1992), p. 35.

2. Iain and Oria Douglas-Hamilton, *Battle for the Elephants* (New York: Viking Penguin, 1992), p. 35.

3. *Ibid.,* p. 37.

4. *Ibid.,* p. 33.

5. *Ibid.,* p. 38.

6. Cynthia Moss, *Elephant Memories* (New York: William Morrow and Company, 1988), pp. 50–51.

CHAPTER 12 **An Accident of History**

1. Charles Darwin, *op. cit.,* p. 459.

2. Robert Broom, *The Coming of Man: Was It Accident or Design?,* p. 220.

3. Stephen Jay Gould, "The evolution of life on the Earth," *Scientific American,* October 1994, p. 89.

4. David M. Raup, "Geological crises in the geologic past," in *Biodiversity,* E. O. Wilson, ed., p. 55.

CHAPTER 13   The Sixth Extinction

1. Edward O. Wilson, "Is humanity suicidal?" *New York Times Magazine*, 30 May 1993, p. 26.

2. Les Kaufman, "Why the ark is sinking," in *The Last Extinction*, Les Kaufman and Kenneth Mallory, eds. (Cambridge: MIT Press, 1993), p. 43.

3. Charles C. Mann, "Extinction: are ecologists crying wolf?" *Science*, vol. 253 (1991), pp. 736–738.

4. Julian L. Simon, "Disappearing species, deforestation, and data," *New Scientist*, 15 May 1986, p. 60.

5. Julian Simon, in Norman Myers and Julian L. Simon, *Scarcity or Abundance?* (New York: W. W. Norton, 1994), p. 35.

6. Julian L. Simon and Aaron Wildavsky, "Facts, not species, are imperiled," *New York Times*, 13 May 1993, p. 23.

7. Thomas E. Lovejoy, "Species leave the ark one by one," in *The Preservation of Species*, Brian Norton, ed. (Princeton: Princeton University Press, 1986), p. 14.

8. David Skole and Compton Tucker, "Tropical deforestation and habitat fragmentation in the Amazon," *Science*, vol. 260 (1993), p. 1909.

9. Paul R. Ehrlich and Anne H. Ehrlich, "The value of diversity," *Ambio*, vol. 21 (1992), p. 225.

10. Julian Simon, in Norman Myers and Julian L. Simon, *Scarcity or Abundance?*, p. 65.

11. Edward O. Wilson, *The Diversity of Life*, p. 220.

12. Julian Simon, cited in William K. Stevens, "Species loss: crisis or false alarm?" *New York Times*, 20 August 1991, p. C8.

13. Julian Simon, in Norman Myers and Julian L. Simon, *Scarcity or Abundance?*, p. 35.

14. Thomas E. Lovejoy, "Species leave the ark one by one," in *The Preservation of Species*, Brian Norton, ed., p. 14.

15. Stuart Pimm, "So many species in so little time," *The Sciences* (in press).

16. *Ibid.*

17. M. P. Nott *et al.*, "Modern extinctions in the kilo-death range," *Current Biology*, vol. 5 (1995), p. 3.

18. Stuart Pimm, "So many species in so little time." Manuscript.

19. Daniel Simberloff, "The ecology of extinction," *Acta Palaeontologica Polonica*, vol. 38 (1994), p. 168.

20. *Ibid.*, p. 171.

CHAPTER 14   Does It Matter?

1. Paul R. Ehrlich, "Population diversity and the future of ecosystems," *Science*, vol. 254 (1991), p. 175.

2. Julian Simon, in Norman Myers and Julian L. Simon, *Scarcity or Abundance?*, p. 35.

3. *Ibid.*

4. Les Kaufman, "Why the ark is sinking," in *The Last Extinction,* Les Kaufman and Kenneth Mallory, eds., p. 2.

5. Michael D. Copeland, "No red squirrels? Mother Nature may be better off," *Wall Street Journal,* 7 June 1990.

6. Stephen Jay Gould, "The golden rule—a proper scale for our environmental crisis," *Natural History,* September 1990, p. 30.

7. *Ibid.*

8. David M. Raup, *Extinction: bad genes or bad luck?,* p. 17.

Index

Italicized page numbers refer to illustrations

ABOUT THE AUTHORS

RICHARD LEAKEY is the world's most famous living paleoan-thropologist. For five years, he was the director of Kenya's Wild-life Services, where he was successful in bringing elephant poaching to a halt. He lives in Kenya.

ROGER LEWIN is a respected author in his own right. The Royal Society of London judged his book *Bones of Contention* to be the best science book published in Great Britain in 1988. He lives in Cambridge, Massachusetts.